the medical

masquerade

Copyright ©2025 by Clayton J. Baker, M.D.

All rights reserved. This book or any portion thereof may not be reproduced or used in any manner whatsoever without the express written permission of the publisher except for the use of brief quotations in a book review.

ISBN Print: 9781630692957
ISBN Ebook: 9781630692919

Published by Brownstone Institute 2025
Creative Commons Attribution 4.0 International

Brownstone Institute
Austin, Texas
2025

the medical masquerade

A PHYSICIAN EXPOSES THE
DECEPTIONS OF COVID

CLAYTON J. BAKER, M.D.

BROWNSTONE
INSTITUTE

to my family

CONTENTS

Foreword by Jeffrey Tucker	ix
Introduction	1
The Four Pillars of Medical Ethics Were Destroyed in the Covid Response	11
The Pharmacological Path to Soft-Core Totalitarianism	29
Solutions to Vaccine Troubles in Ten Sentences	35
The Dirty Secret About How Masks Really "Work"	45
My Golden Retriever Confronts the Medical Juggernaut	49
The Depopulation Bomb: A Halloween Sci-Fi Tale	55
What is Medical Freedom, Exactly?	65
Ten New Year's Resolutions to Restore Medical Freedom	77
Medicine Has Been Fully Militarized	91
France Teeters on the Brink	97
Why Are Health Care Students Still Forced to Get Covid-19 Boosters?	103
The Crucifixion of Kulvinder Kaur	109
Crush the Flu d'État!	115
Does "One Health" include assassinations?	121
Questioning Modern Injection Norms	133
Pandemic Preparedness: Arsonists Run the Fire Department	141
Open Letter to Students and Parents about Vaccines	157
Could Bird Flu Be the October Surprise?	167

Monkeypox: Evidence of the "Pandemic Preparedness" Lie	171
Six Simple Steps to Pharma Reform	181
The Pandemic Planners Come for Hoof and Hen… and Us Again	193
Caligula's Horse and the US Senate	199
What is Joe Biden's Life Expectancy?	203
Disaster: CDC Calls for Covid-19 Vaccines on the Child Immunization Schedule	209
A Voter's Primer: The Seven Health Policy Habits of Insanely Progressive People	215
Covid-19 is Becoming Milder, but the Left Stays Toxic as Ever	221
The Top 10 Covid Villains of 2021	225
Covid-19 in 10 Sentences	231
No, seriously: Outgoing NIH director picks up guitar and sings 'Covid over the Rainbow'	235
The Madness Stops only when Fauci is Stopped	239
Hey, Francis Collins: If you've been at the NIH too long, then what about Anthony Fauci?	241
Fauci should 'get' while the gettin' is good	245
The New Law	249
Appendix A	251
Appendix B	255
Acknowledgements	263
About Brownstone Institute	265
Index	267

FOREWORD

by Jeffrey Tucker

At first it seemed like a calamitous error in the deployment of public health measures. We were locked down, threatened with the ability even to hop in our cars and drive to the next state. Vague orders were coming down from somewhere that if we did that we would have to quarantine for two weeks on each side of the border. Then we were told not to hold any gatherings in our houses. It was only for two weeks but I was incredulous. Exactly what were we trying to achieve here?

I went for a drive. At the time I had a two-seater convertible that made too much noise. I was wearing a suit, scarf, and hat, and rocked up to my favorite distillery. The hipster who usually explained the vanilla notes of their stock of bourbon had a different countenance entirely. She was dressed in a garb of woe, and selling hand sanitizer.

I cracked up, entirely inappropriately, and then asked for 20 bottles just to have as souvenirs of this insanity. She grew furious and denounced me for going on a "joy ride" and attempting to buy up supplies of sanitizer from people who needed it. I cannot remember experiencing such disapproval. I said: "You are serious, aren't you?"

"Very," she responded.

Yikes. So I got back in my car, wondering what in the world had gone wrong in the world. There was said to be a virus loose on the land. But there is always a virus loose on the land, millions and billions of them, but this one was said to be more vicious than ever. However, I had seen the data and knew the demographics. I also knew that this wave would end precisely how they had always ended, with herd immunity from exposure. Such is the delicate dance we all perform daily with the microbial kingdom.

This stupidity will end in a fortnight, I kept telling myself, and then everyone will laugh, learn a lesson, and move on. But that did not happen. It went on and on, with restrictions tightening by the day and ever more craziness, including drones flying overhead to ferret out house parties and funerals and report them to the local media, which became part of the state enterprise.

Later I got a phone call from a guy who worked with George W. Bush on bioweapons. He explained that the lockdowns were good because this way we could wait for the vaccine. I laughed and said this was preposterous because nothing effective could be developed so quickly if ever. He assured me otherwise and hung up. I completely dismissed the possibility that this was even real.

The months rocked on and on, all the way to November, when of course most people had to vote remotely to avoid infection. I myself was offered many ballots as I travelled around the country. I swear that I could have voted 5 times. Apparently millions did, so far as we can tell.

Within a year, Brownstone Institute had been founded and finally I was in contact with people such as Dr. Clayton Baker, who knew that the real story went far beyond a bonkers public health response. The military was involved along with the intelligence services, not just on a national level but on a global one. This was not a mistake but rather a genuine coup d'etat against civilian government in favor of a cabal that operated mostly in secret. Not being accustomed to thinking this way, I could hardly wrap my brain around it.

Years have passed. I'm a different person. Like everyone else. Our old networks collapsed and so did the institutions we once trusted. We inhabit different social spaces now, and a different frame of mind. Now we know things, such as that many mainstream publications, institutions, funding sources, and even retail outlets are nothing more or less than arms of the national security state. It's all still invisible to most but it is now very visible to us because our times have trained us, as if forged by fire.

Dr. Baker is a rare case, an Ivy-trained medical doctor who saw through the hoax and racket from the beginning. He was there every step of the way, calling it out, speaking truth to power, and risking everything to go against the mightiest powers in the world. I'm happy that he came to call Brownstone home for his writing and speaking. In a few years, he will be recognized as the prophet he is. You will soon agree once you start and finish his collection of essays.

How close are we to the point that the public mind comes around to realizing just how hard we were all trolled? I'm not sure but we are closer to that point now than we were a few years ago. We might be many years away before the full truth dawns.

I share with Dr. Baker the passionate desire to stop official history from telling of our times that a killer virus nearly wiped out humanity but for the salvific efforts of pharmaceutical companies. There is not a shred of truth in that claim, as he shows. It was not just a mistake. It was far worse, far more insidious.

The pathetic young women selling sanitizer will likely never come to terms with it. Most people will not. But you hold this book, so you are in a position to know the truth. You can handle it. We all can. Thank you, Dr. Baker for dedicating yourself to finding it and sharing it.

Introduction

*It is better to be unhappy and know the worst,
than to be happy in a fool's paradise.*

FYODOR DOSTOEVSKY

Did the world change because of Covid, or did we?

As I review this volume of essays, all written since the lockdowns began in March 2020, this question keeps coming to mind.

Since Covid, the world seems different. My own attempt to understand how and why it all happened took me step by step into the labyrinth of lies, corruption, and malice that lay behind the lockdowns, the assault on civil rights, the generational suffering, and the countless deaths of the Covid era. With almost every step the way became a bit darker.

On a bad day, I see no end to the human potential for wickedness, especially in those who seek and hold power. The more one learns about the likes of Anthony Fauci, Bill Gates, Tedros Ghebreyesus, Klaus Schwab, and their like, the harder it is to feel otherwise.

On a bad day, I cannot comprehend the credulity and carelessness of so many people. It seems that all tyrants need to do is inject some collective fear, and the public becomes incapable of critical thought, frank speech, or resistance to the most wanton abuse. All that a great many people can muster the nerve to do under such circumstances, it seems, is to turn on the few among them who do manage to resist.

Fortunately there are good days, too.

On a good day, I conclude that a great portion of the world has come to realize, at least intuitively, that they were hoodwinked during Covid, that the whole event was a lie and an act of tyranny. I believe that enough eyes have been opened to stop it from happening again.

On a good day, I remember that because of Covid, I have gotten to know many intelligent, courageous, and truly humane people, probably none of whom I would have met otherwise. Many of these people have risked more, lost more, and accomplished more than me. Sirach teaches that when you meet the wise, your feet should wear away their doorstep. I've had the good fortune to communicate and even collaborate with many of them.

These good and excellent people – the ones who most actively resisted the evil that lay behind Covid – provide hope. In fact, they may be our best hope. They have been persecuted, silenced, canceled, fired, de-platformed, de-licensed, de-monetized, arrested, and some even imprisoned.

But they have not been destroyed.

They are still standing, still speaking out, still fighting for what is true, just, and good. They are still striving to preserve the dignity and freedom of their fellow human beings, including those who still resent them, or even hate them. They have grown in influence and public acceptance, and rightfully so.

Furthermore, as a result of the gradual exposure of the lies, gaslighting, and psyops to which ordinary citizens were subjected during Covid, the *modus operandi* of our governments, intelligence agencies, militaries, corporations, and so-called 'elites' have been exposed.

Another positive if unforeseeable result is that long-term dissidents, truth tellers, and whistleblowers who were marginalized and persecuted for decades are now finally receiving renewed attention.

True heroes like Julian Assange, Edward Snowden, Andrew Wakefield, Meryl Nass, Dane Wigington, and others, long ago recognized and began the fight against the civilizational and governmental corruption that made the Covid catastrophe possible. Many of them were doing so decades before the arrival of Covid-era dissidents like myself.

All of these people paid dearly for their prescience, courage, and stubborn effort to reveal the illegal, immoral, and even murderous nature of our governments and institutions. Some of them paid nearly everything. But now the world is beginning to see these people anew, and it is beginning to take their messages seriously.

This provides even greater hope. And hope is, after all, along with faith and love, one of the three things that abide.

Growth and progress toward the good require change. Change is usually difficult and often painful. This does not make it less necessary.

Like many people who were awakened, red-pilled, activated, or even radicalized by Covid (and I have been called all of those things), I have lost some friends. In some cases I have been rejected. In others, I have consciously reduced the time I spend with certain people. At first this saddened me. Now I think it probably cannot be otherwise.

Once again, has the world changed, or have we?

Covid taught me that dissidents cannot simply pick and choose their colleagues. Once you become an adversary to the existing power structure, you're on your own, pal. There may be friends for you out there, but they're isolated just like you. You find allies one at a time.

Where do you find them? In places you never went before you became an outsider: at street corner protests, in heavily censored social media groups, and as plaintiffs to lawsuits against your own school district.

This re-sorting process is confusing, tiring, and distressing, but it has to happen. Every dissident must go through a process of questioning, reappraisal, and rejection. This process is a two-way street. A dissident rejects the prevailing narrative as false. In return, the conforming majority rejects the dissident as a threat to the established order. From their respective points of view, both sides are correct.

Once the mainstream-citizen-turned-dissident runs this gauntlet, where does he end up? Where he never thought he'd be: with the other malcontents and nonconformists. At a street corner protest, in a heavily censored social media group, or suing his own school district.

The outsiders start working together, and if they stay at it, they may grow in influence and effectiveness. Why?

In the case of the Covid dissidents, our effectiveness grew in large part because we exposed lies, and we refused to stop exposing lies. Maybe it's true that a lie can get halfway around the world before the truth can put its pants on. Over the long term, however, the lie is going to get caught with its pants down a whole lot more often. Point out the lies, keep pointing out the lies, explain why the powers-that-be are telling the lies, and eventually more and more people see through the lies.

The virus came from the wet market, not the lab. A lie.

Two weeks to flatten the curve. A lie.

Six feet to stop the spread. A lie.
Safe and effective. A lie.
Etcetera, etcetera.

Our effectiveness grew because we sought the truth. I believe that deep down most people do hunger for the truth, even if superficially they fear it. Our audience grew because we frankly described, stubbornly investigated, and earnestly interpreted the Covid catastrophe to the best of our ability (see the essay "Covid-19 in Ten Sentences"). Over time, while the legacy media continued to pour out increasingly obvious propaganda, we peeled away at the layers of deceit to reveal just how false and malicious the operation was. Gradually, people listened.

As Covid began to recede, most people longed just to return to (relatively) normal life. However, many of us who ran the risk of taking action and speaking out – and paid a price for doing so – have not let things go. Whether the world changed because of Covid or not, it appears we have.

For me, Covid tore the veneer off almost every institution in life. As a physician, the scales especially fell from my eyes regarding modern medicine. Covid prompted me to weigh my profession on the scales, and it was found wanting.

Prior to Covid, I had taught medical humanities and bioethics for years, both at the bedside and in the classroom. I took medical ethics seriously, and I assumed my profession did too. During Covid, I was appalled at the casual manner in which the fundamental ethical tenets of medicine were cast aside. The entire management level of my profession acted as though patient autonomy was simply null and void. They behaved as though they no longer needed to even consider beneficence, non-maleficence, or justice when caring for patients.

In the essay "The Four Pillars of Medical Ethics Were Destroyed in the Covid Response," I explored this failure of my profession, unsure of how far it would lead. I performed a detailed investigation to determine how many of the key tenets and specific rules of medical ethics had been broken, abused, or ignored during Covid. Almost five thousand words and dozens of references later, I had my answer: all of them. Every one. During Covid, my profession broke all its own ethical rules.

This kind of realization can make one bitter. In fact, bitterness seems to be an occupational hazard of being a dissident. But like envy, bitterness is always ignoble and should be avoided. The best antidote to bitterness is humor, and the child of the two is sarcasm.

To quote Dostoevsky again, *sarcasm is the last refuge of a decent person when the privacy of their soul has been brutally invaded.* Is there a better description of what happened during Covid than that the privacy of our souls was brutally invaded?

Humor generally improves writing. Humor in writing is like beauty in a woman: it's not quite enough all by itself, but it definitely helps. And humor, even sarcastic humor, can help deliver painful news (see "The Top 10 Covid Villains of 2021").

At one point, my editor at Brownstone Institute, Jeffrey Tucker, dropped the hint that he was looking for something a bit lighter in tone than the usually dead-serious material he was publishing. I produced an essay for him entitled "My Golden Retriever Faces the Medical Juggernaut."

The flurry of replies I received regarding this piece, intended as a change-of-pace, came as a surprise. Clearly, identifying the similarities (and similar problems) between human and animal medicine in the wake of Covid affected many readers. People deeply love their pets. I believe this is not only because of the companionship and unconditional love that pet owners receive from their pets, but also because of the connection that even the most domesticated animal provides to an earlier, simpler, and more natural era of human existence.

The emails kept coming in about that essay. One noted the affectionate characterization of my dog, another my lampooning of Pfizer CEO and erstwhile veterinarian Albert Bourla, and a third reported that they laughed out loud. Yet another decried the article for desecrating the honor of decent, hardworking veterinarians everywhere.

It is impossible to know which essays will strike a chord with readers. The essays I think are bound to go 'viral' (a term I both use and dislike) typically don't, while the ones I have no expectations about sometimes take off.

I remember a quote attributed to rock-n-roll musician Alex Chilton. At the tender age of 16, he had a number one hit record. However, after his teens, he never came close to the charts again, despite a long career and ultimate status as one of the classic underground figures of rock-n-roll. Years later, when asked why he hadn't had a hit since he was a teenager, Chilton replied, "my songs sound like hits to me."

Perhaps this is the best approach: write about the issues one thinks are most important, the issues one is most concerned about at present, and the issues for which one believes positive change is possible. Those sound like hits to me.

There is no shortage of material. The societal problems that need examination, elucidation, and exposure are almost endless. Beyond the pharmaceutical-industrial complex, beyond our militarized medical system (see "Medicine Has Been Fully Militarized"), Covid revealed that virtually all of our human institutions are highly susceptible to corruption, and in many cases thoroughly corrupt.

Covid revealed that the institutions that were supposed to provide counterweights to greed, corruption, and power-grabbing

– the press, academia, nonprofit organizations, regulatory agencies, religious institutions, you name it – were in fact captured and complicit with the lies of those in power. We can no longer trust these institutions to be truthful, any more than we can trust Big Pharma, the central banks, or the rapacious, ultra-rich, so-called "elites" such as Bill Gates or the WEF.

Initially during Covid, the foremost task was to stop the obvious civil rights abuses of lockdowns, mandates, and so on. To do so, we had to figure out what was really being done to us, who was behind it, and why they were doing it.

Much of the who/what/where/when/why of the Covid period are now fairly well known to those who have investigated it, although the onion still has unpeeled layers. Many of the underlying mechanisms that facilitated the abuses of Covid have also been identified.

More recently, the focus has turned increasingly to bringing change and reform to these 'mechanisms of harm,' as Lori Weitz has called them. For those fighting for truth and transparency in government, medicine, and industry, as well as for the protection of our fundamental civil rights, we must now, as Bret Weinstein has said, 'play offense.'

Essays in this volume that attempt to take this approach include "Crush the Flu D'état!", "Pandemic Preparedness: Arsonists Run the Fire Department," and "Six Simple Steps to Pharma Reform."

We should also remember that fundamental change for the better must originate from within. We must strengthen our own resolve to never forget what was done to us during Covid, and to never allow it to be done to us again. Any complacency we once held about our existence on Earth should be set aside. We must re-examine our own views of health and medicine ("Questioning Modern Injection Norms,") and rethink our relationship to the collective ("What is Medical Freedom, Exactly?").

So, to answer the question I posed at the beginning of this introduction, I would say the following:

Yes, the world has changed in many ways since March 2020. But much of that apparent change is that the true nature of things has been revealed. And the world needs to change a lot more, especially our human institutions, if we are to prevent the tyranny of Covid from being repeated.

And yes, we have changed in many ways since March 2020 as well. But again, much of that apparent change is that *our* true nature has been revealed. Our complacency, gullibility, dependency, and cowardice, both as individuals and collectively, were mercilessly exploited during Covid. Once again, we need to change ourselves a lot more to prevent it all from being repeated.

To close, I'll quote Dostoevsky one last time: *Anyone who can appease a man's conscience can take away his freedom.* May we never again allow our consciences to be appeased.

CLAYTON J. BAKER, M.D.

Rochester, New York | December 2024

The Four Pillars of Medical Ethics Were Destroyed in the Covid Response

Originally published
May 12, 2023 in *Brownstone Journal*.

Much like a Bill of Rights, a principal function of any Code of Ethics is to set limits, to check the inevitable lust for power, the *libido dominandi*, that human beings tend to demonstrate when they obtain authority and status over others, regardless of the context.

Though it may be difficult to believe in the aftermath of Covid, the medical profession does possess a Code of Ethics. The four fundamental concepts of Medical Ethics – its 4 Pillars – are Autonomy, Beneficence, Non-maleficence, and Justice.

Autonomy, Beneficence, Non-maleficence, and Justice
These ethical concepts are thoroughly established in the profession of medicine. I learned them as a medical student, much as a young Catholic learns the Apostle's Creed. As a medical professor, I

taught them to my students, and I made sure my students knew them. I believed then (and still do) that physicians must know the ethical tenets of their profession, because if they do not know them, they cannot follow them.

These ethical concepts are indeed well-established, but they are more than that. They are also valid, legitimate, and sound. They are based on historical lessons, learned the hard way from past abuses foisted upon unsuspecting and defenseless patients by governments, health care systems, corporations, and doctors. Those painful, shameful lessons arose not only from the actions of rogue states like Nazi Germany, but also from our own United States: witness Project MK-Ultra and the Tuskegee Syphilis Experiment.

The 4 Pillars of Medical Ethics protect patients from abuse. They also allow physicians the moral framework to follow their consciences and exercise their individual judgment – provided, of course, that physicians possess the character to do so. However, like human decency itself, the 4 Pillars were completely disregarded by those in authority during Covid.

The demolition of these core principles was deliberate. It originated at the highest levels of Covid policymaking, which itself had been effectively converted from a public health initiative to a national security/military operation in the United States in March 2020, producing the concomitant shift in ethical standards one would expect from such a change. As we examine the machinations leading to the demise of each of the 4 Pillars of Medical Ethics during Covid, we will define each of these four fundamental tenets, and then discuss how each was abused.

Autonomy

Of the 4 Pillars of Medical Ethics, *autonomy* has historically held pride of place, in large part because respect for the individual patient's autonomy is a necessary component of the other three. Autonomy was the most systemically abused and disregarded of the 4 Pillars during the Covid era.

Autonomy may be defined as the patient's right to self-determination with regard to any and all medical treatment. This ethical principle was clearly stated by Justice Benjamin Cardozo as far back as 1914: "Every human being of adult years and sound mind has a right to determine what shall be done with his own body."

Patient autonomy is "My body, my choice" in its purest form. To be applicable and enforceable in medical practice, it contains several key derivative principles which are quite commonsensical in nature. These include *informed consent, confidentiality, truth-telling,* and *protection against coercion*.

Genuine *informed consent* is a process, considerably more involved than merely signing a permission form. Informed consent requires a *competent* patient, who receives *full disclosure* about a proposed treatment, *understands* it, and *voluntarily* consents to it.

Based on that definition, it becomes immediately obvious to anyone who lived in the United States through the Covid era, that the informed consent process was systematically violated by the Covid response in general, and by the Covid vaccine programs in particular. In fact, every one of the components of genuine informed consent were thrown out when it came to the Covid vaccines:

- Full disclosure about the Covid vaccines – which were extremely new, experimental therapies, using novel technologies, with alarming safety signals from the very start – was systematically denied to the public. Full disclosure was actively suppressed by bogus

anti-"misinformation" campaigns, and replaced with simplistic, false mantras (e.g. "safe and effective") that were in fact just textbook propaganda slogans.
- Blatant coercion (e.g. "Take the shot or you're fired/can't attend college/can't travel") was ubiquitous and replaced voluntary consent.
- Subtler forms of coercion (ranging from cash payments to free beer) were given in exchange for Covid-19 vaccination. Multiple US states held lotteries for Covid-19 vaccine recipients, with up to $5 million in prize money promised in some states.
- Many physicians were presented with financial incentives to vaccinate, sometimes reaching hundreds of dollars per patient. These were combined with career-threatening penalties for questioning the official policies. This corruption severely undermined the informed consent process in doctor-patient interactions.
- Incompetent patients (e.g. countless institutionalized patients) were injected *en masse*, often while forcibly isolated from their designated decision-making family members.

It must be emphasized that under the tendentious, punitive, and coercive conditions of the Covid vaccine campaigns, especially during the "pandemic of the unvaccinated" period, it was virtually impossible for patients to obtain genuine informed consent. This was true for all the above reasons, but most importantly because full disclosure was nearly impossible to obtain.

A small minority of individuals did manage, mostly through their own research, to obtain sufficient information about the Covid-19 vaccines to make a truly informed decision. Ironically, these were principally dissenting healthcare personnel and their families, who, by virtue of discovering the truth, knew "too much." This group overwhelmingly *refused* the mRNA vaccines.

Confidentiality, another key derivative principle of autonomy, was thoroughly ignored during the Covid era. The widespread yet chaotic use of Covid vaccine status as a de facto social credit system, determining one's right of entry into public spaces, restaurants and bars, sporting and entertainment events, and other locations, was unprecedented in our civilization.

Gone were the days when HIPAA laws were taken seriously, where one's health history was one's own business, and where the cavalier use of such information broke Federal law. Suddenly, by extralegal public decree, the individual's health history was public knowledge, to the absurd extent that any security guard or saloon bouncer had the right to question individuals about their personal health status, all on the vague, spurious, and ultimately false grounds that such invasions of privacy promoted "public health."

Truth-telling was completely dispensed with during the Covid era. Official lies were handed down by decree from high-ranking officials such as Anthony Fauci, public health organizations like the CDC, and industry sources, then parroted by regional authorities and local clinical physicians. The lies were legion, and none of them have aged well. Examples include:

- The SARS-CoV-2 virus originated in a wet market, not in a lab
- "Two weeks to flatten the curve"
- Six feet of "social distancing" effectively prevents transmission of the virus
- "A pandemic of the unvaccinated"

- "Safe and effective"
- Masks effectively prevent transmission of the virus
- Children are at serious risk from Covid
- School closures are necessary to prevent spread of the virus
- mRNA vaccines prevent contraction of the virus
- mRNA vaccines prevent transmission of the virus
- mRNA vaccine-induced immunity is superior to natural immunity
- Myocarditis is more common from Covid-19 disease than from mRNA vaccination

It must be emphasized that health authorities pushed deliberate lies, known to be lies at the time by those telling them. Throughout the Covid era, a small but very insistent group of dissenters have constantly presented the authorities with data-driven counterarguments against these lies. The dissenters were consistently met with ruthless treatment of the "quick and devastating takedown" variety now infamously promoted by Fauci and former NIH Director Francis Collins.

Over time, many of the official lies about Covid have been so thoroughly discredited that they are now indefensible. In response, the Covid power brokers, backpedaling furiously, now try to recast their deliberate lies as fog-of-war style mistakes. To gaslight the public, they claim they had no way of knowing they were spouting falsehoods, and that the facts have only now come to light. These, of course, are the same people who ruthlessly suppressed the voices of scientific dissent that presented sound interpretations of the situation in real time.

For example, on March 29, 2021, during the initial campaign for universal Covid vaccination, CDC Director Rochelle Walensky proclaimed on MSNBC that "vaccinated people do not carry the virus" or "get sick," based on both clinical trials and "real-world data." However, testifying before Congress on April 19, 2023, Walensky conceded that those claims are now known to be false, but that this was due to "an evolution of the science." Walensky had the effrontery to claim this before Congress 2 years after the fact, when in actuality, the CDC itself had quietly issued a correction of Walensky's false MSNBC claims back in 2021, a mere 3 days after she had made them.

On May 5, 2023, three weeks after her mendacious testimony to Congress, Walensky announced her resignation.

Truth-telling by physicians is a key component of the informed consent process, and informed consent, in turn, is a key component of patient autonomy. A matrix of deliberate lies, created by authorities at the very top of the Covid medical hierarchy, was projected down the chains of command, and ultimately repeated by individual physicians in their face-to-face interactions with their patients. This process rendered patient autonomy effectively null and void during the Covid era.

Patient autonomy in general, and informed consent in particular, are both impossible where coercion is present. *Protection against coercion* is a principal feature of the informed consent process, and it is a primary consideration in medical research ethics. This is why so-called vulnerable populations such as children, prisoners, and the institutionalized are often afforded extra protections when proposed medical research studies are subjected to institutional review boards.

Coercion not only ran rampant during the Covid era, it was deliberately perpetrated on an industrial scale by governments, the pharmaceutical industry, and the medical establishment. Thousands of American healthcare workers, many of whom had served on the front lines of care during the early days of the pandemic in 2020 (and had already contracted Covid-19 and developed natural immunity) were fired from their jobs in 2021 and 2022 after refusing mRNA vaccines they knew they didn't need, would not consent to, and yet for which they were denied exemptions. "Take this shot or you're fired" is coercion of the highest order.

Hundreds of thousands of American college students were required to get the Covid shots and boosters to attend school during the Covid era. These adolescents, like young children, have statistically near-zero chance of death from Covid-19. However, they (especially males) are at statistically highest risk of Covid-19 mRNA vaccine-related myocarditis.

According to the advocacy group nocollegemendates.com, as of May 2, 2023, approximately 325 private and public colleges and universities in the United States still have active vaccine mandates for students matriculating in the *fall of 2023*. This is true despite the fact that it is now universally accepted that the mRNA vaccines do not stop contraction or transmission of the virus. They have zero public health utility. "Take this shot or you cannot go to school" is coercion of the highest order.

Countless other examples of coercion abound. The travails of the great tennis champion Novak Djokovic, who has been denied entry into both Australia and the United States for multiple Grand Slam tournaments because he refuses the Covid vaccines, illustrate in broad relief the "man without a country" limbo in which the unvaccinated found (and to some extent still find) themselves, due to the rampant coercion of the Covid era.

Beneficence

In medical ethics, *beneficence* means that physicians are obligated to act for the benefit of their patients. This concept distinguishes itself from non-maleficence (see below) in that it is a positive requirement. Put simply, all treatments done to an individual patient should do good to that individual patient. If a procedure cannot help you, then it shouldn't be done to you. In ethical medical practice, there is no "taking one for the team."

By mid-2020 at the latest, it was clear from existing data that SARS-CoV-2 posed truly minimal risk to children of serious injury and death – in fact, the pediatric Infection Fatality Rate of Covid-19 was known in 2020 to be less than half the risk of being struck by lightning. This feature of the disease, known even in its initial and most virulent stages, was a tremendous stroke of pathophysiological good luck, and should have been used to the great advantage of society in general and children in particular.

The opposite occurred. The fact that SARS-CoV-2 causes extremely mild illness in children was systematically hidden or scandalously downplayed by authorities, and subsequent policy went unchallenged by nearly all physicians, to the tremendous detriment of children worldwide.

The frenzied push for and unrestrained use of mRNA vaccines in children and pregnant women – which continues at the time of this writing in the United States – outrageously violates the principle of beneficence. And beyond the Anthony Faucis, Albert Bourlas, and Rochelle Walenskys, thousands of ethically compromised pediatricians bear responsibility for this atrocity.

The mRNA Covid vaccines were – and remain – new, experimental vaccines with zero long-term safety data for either the specific antigen they present (the spike protein) or their novel functional platform (mRNA vaccine technology). Very early on, they were known to be ineffective in stopping contraction or transmission of the virus, rendering them useless as a public health measure. Despite this, the public was barraged with bogus "herd immunity" arguments. Furthermore, these injections displayed alarming safety signals, even during their tiny, methodologically challenged initial clinical trials.

The principle of beneficence was entirely and deliberately ignored when these products were administered willy-nilly to children as young as 6 months, a population to whom they could provide zero benefit – and as it turned out, that they would harm. This represented a classic case of "taking one for the team," an abusive notion that was repeatedly invoked against children during the Covid era, and one that has no place in the ethical practice of medicine.

Children were the population group that was most obviously and egregiously harmed by the abandonment of the principle of beneficence during Covid. However, similar harms occurred due to the senseless push for Covid mRNA vaccination of other groups, such as pregnant women and persons with natural immunity.

Non-Maleficence

Even if, for argument's sake alone, one makes the preposterous assumption that all Covid-era public health measures were implemented with good intentions, the principle of *non-maleficence* was nevertheless broadly ignored during the pandemic. With the growing body of knowledge of the actual motivations behind so many aspects of Covid-era health policy, it becomes clear that non-maleficence was very often replaced with outright malevolence.

In medical ethics, the principle of non-maleficence is closely tied to the universally cited medical dictum of *primum non nocere*, or, "First, do no harm." That phrase is in turn associated with a statement from Hippocrates' *Epidemics*, which states, "As to diseases make a habit of two things – to help, or at least, to do no harm." This quote illustrates the close, bookend-like relationship between the concepts of beneficence ("to help") and non-maleficence ("to do no harm").

In simple terms, non-maleficence means that if a medical intervention is likely to harm you, then it shouldn't be done to you. If the risk/benefit ratio is unfavorable to you (i.e., it is more likely to hurt you then help you), then it shouldn't be done to you. Pediatric Covid mRNA vaccine programs are just one prominent aspect of Covid-era health policy that absolutely violate the principle of non-maleficence.

It has been argued that historical mass-vaccination programs may have violated non-maleficence to some extent, as rare severe and even deadly vaccine reactions did occur in those programs. This argument has been forwarded to defend the methods used to promote the Covid mRNA vaccines. However, important distinctions between past vaccine programs and the Covid mRNA vaccine program must be made.

First, past vaccine-targeted diseases such as polio and smallpox were deadly to children – unlike Covid-19. Second, such past vaccines were effective in both preventing contraction of the disease in individuals and in achieving eradication of the disease – unlike Covid-19. Third, serious vaccine reactions were truly rare with those older, more conventional vaccines – again, unlike Covid-19.

Thus, many past pediatric vaccine programs had the potential to meaningfully benefit their individual recipients. In other words, the *a priori* risk/benefit ratio may have been favorable, even in tragic cases that resulted in vaccine-related deaths. This was never even arguably true with the Covid-19 mRNA vaccines.

Such distinctions possess some subtlety, but they are not so arcane that the physicians dictating Covid policy did not know they were abandoning basic medical ethics standards such as non-maleficence. Indeed, high-ranking medical authorities had ethical consultants readily available to them – witness that Anthony Fauci's *wife*, a former nurse named Christine Grady, served as chief of the Department of Bioethics at the National Institutes of Health Clinical Center, a fact that Fauci flaunted for public relations purposes.

Indeed, much of Covid-19 policy appears to have been driven not just by rejection of non-maleficence, but by outright malevolence. Compromised "in-house" ethicists frequently served as apologists for obviously harmful and ethically bankrupt policies, rather than as checks and balances against ethical abuses.

Schools never should have been closed in early 2020, and they absolutely should have been fully open without restrictions by fall of 2020. Lockdowns of society never should have been instituted, much less extended as long as they were. Sufficient data existed in real time such that both prominent epidemiologists (e.g. the authors of the Great Barrington Declaration) and select individual clinical physicians produced data-driven documents (see Appendix B) publicly proclaiming against lockdowns and school closures by mid-to-late 2020. These were either aggressively suppressed or completely ignored.

Numerous governments imposed prolonged, punishing lockdowns that were without historical precedent, legitimate epidemiological justification, or legal due process. Curiously, many of the worst offenders hailed from the so-called liberal democracies of the Anglosphere, such as New Zealand, Australia, Canada, and deep blue parts of the United States. Public schools in the United States were closed an average of 70 weeks during Covid. This was far longer than most European Union countries, and longer still than Scandinavian countries who, in some cases, never closed schools.

The punitive attitude displayed by health authorities was broadly supported by the medical establishment. The simplistic argument developed that because there was a "pandemic," civil rights could be decreed null and void – or, more accurately, subjected to the whims of public health authorities, no matter how nonsensical those whims may have been. Innumerable cases of sadistic lunacy ensued.

At one point at the height of the pandemic, in this author's locale of Monroe County, New York, an idiotic Health Official decreed that one side of a busy commercial street could be open for business, while the opposite side was closed, because the center of the street divided two townships. One town was code "yellow," the other code "red" for new Covid-19 cases, and thus businesses mere yards from one another survived or faced ruin. Except, of course, the liquor stores, which, being "essential," never closed at all. How many thousands of times was such asinine and arbitrary abuse of power duplicated elsewhere? The world will never know.

Who can forget being forced to wear a mask when walking to and from a restaurant table, then being permitted to remove it once seated? The humorous memes that "you can only catch Covid when standing up" aside, such pseudo-scientific idiocy smacks of totalitarianism rather than public health. It closely mimics the deliberate humiliation of citizens through enforced compliance with patently stupid rules that was such a legendary feature of life in the old Eastern Bloc.

And I write as an American who, while I lived in a deep blue state during Covid, never suffered in the concentration camps for Covid-positive individuals that were established in Australia.

Those who submit to oppression resent no one, not even their oppressors, so much as the braver souls who refuse to surrender. The mere presence of dissenters is a stone in the quisling's shoe – a constant, niggling reminder to the coward of his moral and ethical inadequacy. Human beings, especially those lacking personal integrity, cannot tolerate much cognitive dissonance. And so they turn on those of higher character than themselves.

This explains much of the sadistic streak that so many establishment-obeying physicians and health administrators displayed during Covid. The medical establishment – hospital systems, medical schools, and the doctors employed therein – devolved into a medical Vichy state under the control of the governmental/industrial/public health juggernaut.

These mid- and low-level collaborators actively sought to ruin dissenters' careers with bogus investigations, character assassination, and abuse of licensing and certification board authority. They fired the vaccine refuseniks within their ranks out of spite, self-destructively decimating their own workforces in the process. Most perversely, they denied early, potential life-saving treatment to all their Covid patients. Later, they withheld standard therapies for non-Covid illnesses – up to and including organ transplants – to patients who declined Covid vaccines, all for no legitimate medical reason whatsoever.

This sadistic streak that the medical profession displayed during Covid is reminiscent of the dramatic abuses of Nazi Germany. However, it more closely resembles (and in many ways is an extension of) the subtler yet still malignant approach followed for decades by the United States Government's medical/industrial/public health/national security nexus, as personified by individuals like Anthony Fauci. And it is still going strong in the wake of Covid.

Ultimately, abandonment of the tenet of non-maleficence is inadequate to describe much of the Covid-era behavior of the medical establishment and those who remained obedient to it. Genuine malevolence was very often the order of the day.

Justice

In medical ethics, the Pillar of *justice* refers to the fair and equitable treatment of individuals. As resources are often limited in health care, the focus is typically on *distributive* justice; that is, the fair and equitable allocation of medical resources. Conversely, it is also important to ensure that the burdens of health care are as fairly distributed as possible.

In a just situation, the wealthy and powerful should not have instant access to high-quality care and medicines that are unavailable to the rank and file or the very poor. Conversely, the poor and vulnerable should not unduly bear the burdens of health care, for example, by being disproportionately subjected to experimental research, or by being forced to follow health restrictions to which others are exempt.

Both of these aspects of justice were disregarded during Covid as well. In numerous instances, persons in positions of authority procured preferential treatment for themselves or their family members. Two prominent examples:

According to ABC News, "in the early days of the pandemic, New York Governor Andrew Cuomo prioritized COVID-19 testing for relatives including his brother, mother and at least one of his sisters, when testing wasn't widely available to the public." Reportedly, "Cuomo allegedly also gave politicians, celebrities and media personalities access to tests."

In March 2020, Pennsylvania Health Secretary Rachel Levine directed nursing homes to accept Covid-positive patients, despite warnings against this by trade groups. That directive and others like it subsequently cost tens of thousands of lives. Less than two months later, Levine confirmed that her own 95 year-old mother had been removed from a nursing home to private care. Levine was subsequently promoted to 4-star Admiral in the US Public

Health Service by the Biden administration.

The burdens of lockdowns were distributed extremely unjustly during Covid. While average citizens remained in lockdown, suffering personal isolation, forbidden to earn a living, the powerful flouted their own rules. Who can forget how US House Speaker Nancy Pelosi broke the strict California lockdowns to get her hair styled, or how British Prime Minister Boris Johnson defied his own supposedly life-or-death orders by throwing at least a dozen parties at 10 Downing Street in 2020 alone? House arrest for thee, wine and cheese for me.

But California Governor Gavin Newsom might take the cake. At first glance, given both his BoJo-esque, lockdown-defying dinner with lobbyists at the ultra-swanky Napa Valley restaurant The French Laundry, and his decision to send his own children to expensive private schools which were fully open for 5-day in-school learning during the prolonged California school closures, one might think of Newsom as a Covid-era Robin Hood. That is, until one realizes that he presided over those same punishing, inhumane lockdowns and school closures. He was actually the Sheriff of Nottingham.

To a decent person with a functioning conscience, this level of sociopathy is difficult to comprehend. What is crystal clear is that anyone capable of the hypocrisy that Gavin Newsom displayed during Covid should not be anywhere near a position of power in any society.

Two additional points should be emphasized. First, these egregious acts were rarely, if ever, called out by the medical establishment. Second, the behaviors themselves show that those in power never truly believed their own narrative. Both the medical establishment and the power brokers knew the danger posed by the virus, while real, was grossly overstated. They knew the lockdowns, social distancing, and masking of the population at large were kabuki theater at best, and soft-core totalitarianism at worst. The lockdowns were based on a gigantic lie, one they neither believed nor felt compelled to follow themselves.

Solutions and Reform

The abandonment of the 4 Pillars of Medical Ethics during Covid has contributed greatly to an historic erosion of public trust in the healthcare industry. This distrust is entirely understandable and richly deserved, however harmful it may prove to be for patients. For example, at a population level, trust in vaccines *in general* has dramatically reduced worldwide, compared to the pre-Covid era. Millions of children now stand at increased risk from proven vaccine-preventable diseases due to the thoroughly unethical push for unnecessary, indeed harmful, universal Covid-19 mRNA vaccination of children.

Systemically, the medical profession desperately needs ethical reform in the wake of Covid. Ideally, this would begin with a strong reassertion of and recommitment to the 4 Pillars of Medical Ethics, again with patient autonomy at the forefront. It would continue with prosecution and punishment of those individuals most responsible for the ethical failures, from the likes of Anthony Fauci on down. Human nature is such that if no sufficient deterrent to evil is established, evil will be perpetuated.

Unfortunately, within the medical establishment, there does not appear to be any impetus toward acknowledgement of the profession's ethical failures during Covid, much less toward true reform. This is largely because the same financial, administrative, and regulatory forces that drove Covid-era failures remain in control of the profession. These forces deliberately ignore the catastrophic harms of Covid policy, instead viewing the era as a sort of test run for a future of highly profitable, tightly regulated health care. They view the entire Covid-era martial-law-as-public-health approach as a prototype, rather than a failed model.

Reform of medicine, if it happens, will likely arise from individuals who refuse to participate in the "Big Medicine" vision of health care. In the near future, this will likely result in a fragmentation of the industry analogous to that seen in many other aspects of post-Covid society. In other words, there is apt to be a "Great Re-Sort" in medicine as well.

Individual patients can and must affect change. They must replace the betrayed trust they once held in the public health establishment and the healthcare industry with a critical, *caveat emptor*, consumer-based approach to their health care. If physicians were ever inherently trustworthy, the Covid era has shown that they no longer are so.

Patients should become highly proactive in researching which tests, medications, and therapies they accept for themselves (and especially for their children). They should be unabashed in asking their physicians for their views on patient autonomy, mandated care, and the extent to which their physicians are willing to think and act according to their own consciences. They should vote with their feet when unacceptable answers are given. They must learn to think for themselves and ask for what they want. And they must learn to say no.

The Pharmacological Path to Soft-Core Totalitarianism

Originally published June 15, 2023 in *Brownstone Journal*.
Also published in the July 2024 issue of the Eagle Forum Report
founded by Phyllis Schlafly *(Volume 8/Number 7)*.

In my recent article on the destruction of medical ethics during the Covid era, I used a term that generated some unanticipated feedback. What exactly did I mean by "soft-core totalitarianism?" Was it a real concept, or just a turn of phrase?

It's a concept all right, and I believe it aptly describes the sociopolitical – or is it sociopathological? – condition in which we find ourselves in the post-Covid world, both in the United States and throughout the other Western so-called liberal democracies. It certainly seems to describe the direction in which our collective culture is heading.

Let's define *soft-core totalitarianism* as a political system, characterized by two features:

- First, there is centralized, autocratic control of the state by the executive branch of the government and its cronies (the *totalitarianism* part).
- Second, *by design*, the offenses to civil liberties are applied subtly and gradually enough, the standard of living is kept high enough, and a sufficient menu of seductive distractions are intentionally provided, that most individuals either don't object, or don't even notice (the *soft-core* part).

Soft-core totalitarianism can be readily contrasted with its hard-core counterpart, recent examples of which would be the killing fields of Khmer Rouge-era Cambodia, or the enforced starvation conditions of contemporary North Korea.

As with soft-core pornography, this comparison favorably disposes soft-core totalitarianism to the more pliant and undiscerning among us, who might say, "Well, yes, I suppose it isn't ideal, but at least it's not that horrible hard-core stuff!"

Also like soft-core pornography, soft-core totalitarianism is designed to possess a titillating and seductive quality that opposes one's better moral judgment. To the weak-willed and soft-headed, the statement "You'll own nothing and be happy" holds a similar appeal to "All the classiest girls, ready to talk about your deepest fantasies, just a phone call away."

Soft-core totalitarianism was predicted by past seers who tried to warn us about where Western Civilization was headed. The leavening of tyranny with the intentional supply of banal distractions, creature comforts, and legalized drugs pepper their descriptions. They repeatedly describe a kind of semi-anesthetized, semi-tolerable dystopia:

"Films, football, beer, and above all, gambling filled up the horizon of their minds. To keep them in control was not difficult." ~ **George Orwell**

"There will be, in the next generation or so, a pharmacological method of making people love their servitude, and producing dictatorship without tears, so to speak, producing a kind of painless concentration camp for entire societies, so that people will in fact have their liberties taken away from them, but will rather enjoy it, because they will be distracted from any desire to rebel." ~ **attributed to Aldous Huxley**

And of course, the granddaddy of them all:

"Give them bread and circuses, and they will never revolt." ~ **Juvenal**

Ring any bells? If not, consider the following, all of which occurred contemporaneously to the Covid-era lockdowns, school closures, travel restrictions, mask and vaccine mandates and other assorted assaults on our civil rights:

- Liquor stores were deemed essential from the beginning and remained open for business throughout the entire duration of the Covid lockdowns.
- No fewer than 17 US States legalized recreational marijuana use since the Covid lockdowns began in March 2020.
- 33 US states have opened legalized sports betting markets since the Supreme Court struck down the Professional and

Amateur Sports Protection Act (PASPA) in May 2018.
- Over $220 billion has been wagered in sportsbooks in the US since then.
- The US Border Patrol seized over 15,000 pounds of fentanyl at the Mexican border in 2022 – three times as much as in 2020.
- 110,000 Americans died of drug overdoses in 2022.

Still skeptical? Let me put it another way. I am no Orwell or Huxley (and certainly no Juvenal), but please indulge me, Dear Reader, as I offer you my own soft-core totalitarian dystopian narrative:

As it happened, it was the Ides of March. Clinton Barker's wife shook him awake. It was five-thirty, half an hour before his alarm was set to go off. "Come downstairs," she said, "you've got to hear what they're saying on television."

A talking head on the screen was spouting instructions, the likes of which Clinton had never heard before.

"Everyone is ordered to remain in their homes," the talking head announced. "I repeat: remain in your homes. Schools are closed until further notice. Workplaces are closed, except for those deemed essential by the authorities."

Clinton Barker's three young children, also glued to the television screen, howled with delight. "No school!" they cheered in unison. "NO SCHOOL!"

Clinton's wife shushed them to silence.

"Authorities are insisting: we all have to do our part," the cypher in the suit continued. "If we all stay in our homes for just two weeks – two weeks to flatten the curve, two weeks to stop the spread of the virus – then all will be well.

Do your part and stay in your homes. Stay home with your families, your children. Think of it as a vacation. Relax and spend time with the family. Kick back and check out what's on Webflix. In fact, Stephanie has some wonderful recommendations. Stephanie?"

An attractive blonde woman with a pneumatic figure, sitting next to the cypher, chimed in.

"So many great shows to watch on Webflix these days, Bill," the blonde chirped, her bosom veritably heaving. "There's a fantastic new documentary on wild animal breeders called The Liger King. *You have to see it to believe it. There's also a thrilling new family adventure series, about a group of treasure-hunting teens, called* Barrier Reefs. *The whole family will love it. Plus there's just so many great established series out there...no one could possibly be caught up on all of them:* Tournament of Kings. The Contraltos. Turning Evil. *You name it! As for me, Bill, I'm going straight home after this and I'm just going to binge."*

Clinton Barker's mind raced. What is happening? What if they lock us down longer than they claim? How will we feed the children? How will we pay the rent? How will we...

As weeks passed, Clinton's first fear was indeed realized – the lockdowns persisted, much longer than initially claimed. But all around him, no one seemed to mind very much. The authorities made sure of that. The televisions ran nonstop, and just like the pneumatic blonde, everyone Clinton spoke to on the phone seemed to be binging on streaming entertainment. The grocery stores remained open, and of course, so did the liquor stores.

Before long, the state announced that marijuana had been legalized. Soon thereafter, Clinton received a check from the government, sufficient to cover a few months' groceries. Even the rent was not a problem. The state announced a moratorium on all rent collection until further notice.

Sound familiar?

For those of you still skeptical of my thesis, save your time and read no further. I wish you good luck; I suspect you will need it. To paraphrase Sam Adams, may your chains set lightly upon you, and may posterity forget that you were my countrymen.

For those of you who think I may be on to something, who agree that we are indeed living in an age of soft-core totalitarianism (and which, I fear, is the inevitable gateway to the hard-core variety), I recommend the following potential remedies. I am eager to hear yours as well.

- We must fight tooth and nail against every incursion on civil rights and liberties, regardless of how it affects our creature comforts or our immediate standard of living.
- We must insist that every government action be legal (i.e. passed by law through legislature, and challenged in court if unconstitutional).
- We must fight against executive orders run amok and declarations of emergency called at the drop of a hat. These extralegal abuses must be relegated to the same dustbin of historical irrelevance as papal bulls.
- We must hold accountable those individuals, *especially those who are unelected*, who are guilty of crimes against liberty, civil rights, and humanity. If individuals are not appropriately punished, systems will continue to decay.

And may God have mercy on us all.

Solutions to Vaccine Troubles in Ten Sentences

Originally published July 24, 2023 in Brownstone Journal *and in* The Defender *of Children's Health Defense.*

The uncritical, blind faith in vaccines is the preeminent sacred cow of modern medicine. (It happens to be its preeminent cash cow as well.) It is a quasi-religious, dogmatic article of conviction, rather than a sound scientific theory or an empirically-based clinical precept.

Vaccines have been controversial since their introduction centuries ago. Only in very recent history has there been a rigidly enforced orthodoxy of belief within the medical establishment that vaccines must be unanimously regarded as "safe and effective," no questions asked.

Even more recent is the practice of smearing and labeling anyone questioning this doctrine as a heretic: an "anti-vaxxer." In fact, according to the Merriam-Webster Dictionary, the earliest known use of that now-ubiquitous epithet was only in 2001.

Religious faith has tremendous potential for good in society, but when it is misrepresented as science, its track record is miserable and deadly. "Safe and effective" is not scientific shorthand, or even an advertising slogan; it is a mantra. "Anti-vaxxer" is not a category of person, it is a charge of heresy. And just as vaccine critics are heretics, so the high priests of vaccines, the Faucis of the world, the people who in their own words "represent science," are fanatics.

Does that really sound like science to you? Galileo, Semmelweis, and a few others might disagree.

Any honest person who lived through the Covid-19 era in the United States will acknowledge that the Department of Health and Human Services (HHS) with its lengthy "alphabet soup" of agencies (CDC, NIH (with its NIAID), FDA (with its CBER), etc., etc.), promoted and repeated the "safe and effective" mantra regarding the Covid-19 vaccines throughout an era of intense public fear.

Any honest person will also acknowledge that the mainstream media avidly repeated and amplified the "safe and effective" mantra and stoked the fear, all while ruthlessly attacking anyone questioning that same dogma, labeling them "anti-vaxxers," or sometimes even "murderers."

Little to no mention was made – or allowed – of the gigantic financial incentives and other entanglements these powerful entities have with the vaccine manufacturers, nor the trillions of dollars involved.

Religious dogmas, especially those relentlessly inculcated by powerful forces under extreme conditions, are hard to break free from.

To readers who may know people caught in the rigid, dogmatic belief in the infallibility of vaccines, I offer the following 10 sentences.

Share them with friends, family, and colleagues who cannot seem to reconsider vaccine dogma, especially those with an uncritical view of the current vaccine schedules. Ask them to carefully read each of the 10 sentences below, one at a time, and ask themselves: does this sentence seem true or false to me? If it seems false, *on what basis* do I think it is false? Then move on to the next one and do the same.

(Some of the sentences are complex, but I am confident an intelligent layperson can understand them all.)

When they are finished with all 10 sentences, encourage your friends to ask themselves:

- Do they truly believe that every child in the United States should receive 20 or more different vaccines before age 18?
- Should vaccines ever be mandated?
- Shouldn't we, as an educated, free society, systematically review the official vaccine recommendations, and, just as we would do with Grandma's overflowing pill box, reduce them to the truly necessary minimum?
- Shouldn't we reassert the autonomy of patients over their own bodies?

Here is the trouble with vaccines, in 10 sentences:

1. Like "antibiotics," "vaccines" are a large and diverse class of medicines, and as with all large classes of medicines, different products in the class work by different mechanisms, some being quite effective while others are ineffective, some being reasonably safe for appropriate human use while others are fraught with side effects and toxicities, and therefore to assume that

any large class of medicines – including vaccines – is categorically "safe and effective," is naïve, illogical, false, and dangerous.

2. While the full extent of vaccine toxicity is undetermined, it is a historical fact that numerous vaccines have been proven to be highly toxic and even deadly to patients, via multiple pathophysiological mechanisms, including: a) direct contamination of the vaccine (e.g. the Cutter Incident), b) disease caused by unintended, pathological immune response to the vaccine (e.g. Guillain–Barré syndrome caused by the swine flu vaccine), c) unintended contraction and/or transmission of the disease the vaccine was designed to prevent, *caused by the vaccine itself* (e.g. the current oral polio vaccine), and d) vaccine toxicity of unknown or uncertain cause (e.g. intestinal intussusception with the rotavirus vaccine, and fatal blood clots with the Johnson & Johnson Covid-19 vaccine).

3. In fact, the known toxicity of vaccines is so well-established that a Federal law – the National Childhood Vaccine Injury Act (NCVIA) of 1986 (42 U.S.C. §§ 300aa-1 to 300aa-34) was passed to specifically exempt vaccine manufacturers from product liability, based on the legal principle that vaccines are "unavoidably unsafe" products.

4. Since the 1986 NCVIA act protecting vaccine manufacturers from liability, there has been a dramatic increase in the number of vaccines on the market, as well as the number of vaccines added to the CDC vaccine schedules, with the number of vaccines on the CDC Child and Adolescent schedule rising from 7 in 1986 to 21 in 2023.

5. Of the 21 vaccines on the 2023 CDC Child and Adolescent Immunization Schedule, only a small minority (e.g. measles, mumps, rubella, varicella, and HiB) are capable of providing genuine herd immunity, a fact that negates the common, population-based arguments for mandating the other vaccines, which comprise the sizable majority of the vaccines on the schedule.
6. The pharmaceutical industry has established an almost unimaginable degree of media control, institutional influence, and regulatory capture, via its funding of other entities, as it is a) the largest industry lobby in Washington, DC, b) the second largest industry in TV advertising, c) a major source of personal revenue for high-level HHS "alphabet soup" agency bureaucrats, many of whom hold patent and royalty rights on pharmaceutical products, d) a major funder of influential physician organizations (e.g. the American Academy of Pediatrics and prominent medical journals, and e) involved in payment-based incentivization of practicing physicians, who frequently receive monetary bonuses for high rates of vaccination in their patient panels.
7. The Covid-19 mRNA vaccines were developed and administered to the public a) much faster and with much less testing than any other vaccines on the market, b) under Emergency Use Authorization, c) utilizing a technological platform that had never seen commercial use before, and, despite generating reports of vaccine-related deaths and serious adverse events at much higher rates than traditional vaccines, and despite the fact that they have been removed from the pediatric market in multiple other developed countries, the Covid-19 mRNA

vaccines have already been placed on the CDC Child and Adolescent Immunization Schedule, just a little over 2 years after their introduction to the public.

8. There has been no systematic public accounting by the CDC (or any of the HHS agencies) for the more than 35,000 reported Covid-19 vaccine-related deaths and more than 1,500,000 reported Covid-19 vaccine-related adverse events reported as of July 7, 2023, to the CDC's own Vaccine Adverse Event Reporting System (VAERS), nor for the corresponding numbers of Covid vaccine-related deaths and adverse events reported to Eudravigilance (the European Union's equivalent to VAERS), even as the CDC continues to strongly promote these vaccines for use, including placing them on the CDC Child and Adolescent Immunization Schedule.

9. By labeling the novel Covid mRNA products as "vaccines," the definition of the term "vaccine" has become so broadened that essentially any medication that induces an immune response against a disease may now be dubbed a "vaccine," thereby shielding pharmaceutical companies from liability under the National Childhood Vaccine Injury Act of 1986 to a previously unimagined extent.

10. Vaccine mandates thereby compel citizens to submit to medical treatments a) that are regarded under Federal law to be "unavoidably unsafe," b) that *because* they are unavoidably unsafe, their manufacturers are protected by Federal law from liability for harm done to citizens, c) whose manufacturers and government agencies nevertheless promote publicly as "safe and effective," in direct contradiction to their legal status as "unavoidably unsafe," and d) that have increased tremendously in

number in recent decades, and, with mRNA technology and a broadened definition of the term "vaccine," stand to multiply at an even greater rate in the future.

I hope these 10 sentences will help the unconvinced to reconsider the central dogma surrounding vaccines. We, as a society, need to reject the article of faith that vaccines are fundamentally "safe and effective."

Vaccines, due to their unavoidably unsafe nature, should NEVER be mandated, and a thorough, product-by-product accounting of the individual vaccines needs to be done outside of government agencies.

How can we accomplish this?

Please forgive me if you thought I was done. I have 10 more sentences listing my proposed solutions to the trouble with vaccines. I ask you to trudge through these as well. Most of them are shorter than the first 10. Thank you.

A Proposed Solution to the Trouble with Vaccines in 10 (more) Sentences:

1. The National Childhood Vaccine Injury Act (NCVIA) of 1986 (42 U.S.C. §§ 300aa-1 to 300aa-34) should be repealed, returning vaccines to the same liability status as other drugs.
2. Federal law should be passed prohibiting the mandating of any and all vaccines at all levels of government.
3. Federal law should be passed prohibiting all direct-to-consumer advertising of prescription drugs.
4. Federal law should be passed prohibiting all collaboration between the Department of Health and Human Services' "alphabet soup" agencies (FDA, CDC, NIH, etc.) and

either the Department of Defense (US Army, DARPA, etc.) or the Federal Intelligence Agencies (CIA, DHS, etc.) with regard to vaccine development or vaccine distribution to the public.
5. Federal law should be passed prohibiting all persons working within the HHS agencies from gaining any personal financial benefit from vaccines, including the gaining and holding of patents or royalties, and civil servants in those agencies should be required to take an oath of office not to profit off of any products they approve, regulate, or about which they advise the public.
6. A thorough and public investigation, including criminal prosecutions where appropriate, should be made regarding the key players (both public and private) involved in the development, marketing, manufacture, sale, and administration of the Covid-19 mRNA vaccines, and following the investigation, there should be appropriate reform within the HHS agencies.
7. Detailed, independent, Cochrane-style reviews of every vaccine on the CDC vaccine schedules should be undertaken and made public, and no scientists with financial interests within the pharmaceutical industry should conduct these reviews.
8. Detailed, independent reviews of all reports from the Vaccine Adverse Event Reporting System (VAERS) related to the Covid-19 mRNA vaccines should be undertaken and made public, and appropriate reforms to VAERS should be made.
9. A detailed Congressional review of the money trails related to Covid-era programs, including Operation Warp Speed and the Coronavirus Aid, Relief, and Economic

Security (CARES) Act, should be conducted, focusing on fraud and abuse at all levels, including how private companies such as Pfizer and Moderna profited so enormously from taxpayer-funded initiatives.
10. A open, public discussion and debate should be undertaken on the appropriate role of vaccines in public health, including, among other issues, a) a critical review of the current medical dogma on vaccines, b) an accounting of the mistakes, abuses, and potential lessons of the Covid-19 era, and c) a thorough discussion of the undeniable conflicts between public health as it is now practiced and the fundamental civil rights of citizens.

The medical establishment's current dogma on vaccines ("safe and effective," no questions asked) and its corresponding catechism (the ever-expanding vaccine schedules) are in desperate need of reform. I submit that we begin with the above steps.

Reformers are not heretics, although they are commonly labeled as such by powerful persons resisting reform. I, for one, am not a heretic, nor am I an "anti-vaxxer." I don't want to throw the baby out with the bathwater. The problem is, when one looks closely at the vaccine schedules, there turns out to be a lot more bathwater and a lot less baby than advertised.

It is time for the profession of medicine, and society as a whole, to come out of the Dark Ages on this topic. It is time for an open, forthright reevaluation of vaccines and their role in public health.

The Dirty Secret About How Masks Really "Work"

Originally published on September 3, 2023 *in Brownstone Journal* and on September 11, 2023 in *Epoch Times*.

It is difficult to believe that Public Health™ is trying to force America to mask up again, but here we are.

The question is, why?

The dirty secret is this: Masks don't work by controlling the virus. Masks work by controlling the people.

If we're talking about stopping the spread of the virus, masks simply don't work.

But if we're talking about stoking fear, instilling blind obedience to state authorities, sowing discord between citizens, and publicly "outing" skeptics and dissidents – in other words, creating an authoritarian, even totalitarian system of public health – then masks work very well indeed.

Masks Don't Work at Controlling the Virus

By this late date, it has been established beyond honest scientific doubt that masking is ineffective at stopping the contraction and spread of Covid-19. This is true both at the microscopic level and at the population level.

The early mask mandates regarding Covid-19 were largely "justified" on the assertion that the SARS-CoV-2 virus was not prone to airborne spread. However, the SARS-CoV-2 virus *has* since been proven to be an airborne virus (like influenza), meaning it can remain circulating in room air for extended periods of time, and spreads in this manner. SARS-CoV-2 viruses have also been proven to be much smaller in size than the holes in cloth and surgical masks.

Therefore, at a microscopic level, Harvey Risch is correct: trying to block the SARS-CoV-2 virus with a surgical mask is quite literally like trying to keep mosquitos out of your yard by erecting a chain-link fence.

At a population level, the latest Cochrane meta-analysis of the available randomized, controlled trials surrounding masking and respiratory viruses concluded that "Wearing masks in the community probably makes little or no difference to the outcome of influenza-like illness (ILI)/COVID-19 like illness compared to not wearing masks. Wearing masks in the community probably makes little or no difference to the outcome of laboratory-confirmed influenza/SARS-CoV-2 compared to not wearing masks."

(It should be noted that as the mask debate has been resurrected, Cochrane has been under intense pressure by pro-mask entities to addend and modify their comments about this study, to which the organization has capitulated.)

Furthermore, this study is only one in addition to the hundreds of other studies that clearly outline the epidemiologic ineffectiveness and real harms of masks, many of which have been known since at least 2021.

To summarize: at the microscopic level, masks do not stop the exit or entry of the virus into human bodies, and at the population level, mask use has not been shown to provide any benefit, and has been shown to have numerous harms.

Masks Do Work at Controlling People

The entire Public Health™ enterprise in the West has a strong political and authoritarian impulse built into it from its very conception. While a detailed review of this is beyond the scope of this article, it harkens back at least to the figure of Rudolf Virchow, the preeminent 19th century German physician, opponent of Semmelweis and Darwin, and founder of so-called "social medicine," who famously wrote that "Medicine is a social science, and politics nothing but medicine at a larger scale."

The attitude that Public Health™ should possess the power to dictate national and local political policy for the "public good" (as they, the "experts," unilaterally determine it to be) has increased over the past century, especially in the United States. Around it there have grown vast, lucrative industries, from which (since at least the Bayh-Dole Act), Public Health™ officials often profit handsomely. The vaccine industry is only the most obvious of these.

During the Covid era, the authoritarianism of Public Health™ morphed into totalitarian mode, with the unprecedented lockdowns, school closures, travel restrictions, vaccine mandates, etc. that we all endured. The most visible and most easily enforceable symbol of this power grab were masks.

Masks, even the comically useless ones made of old handkerchiefs, or the filthy, week-old paper surgical ones seen on countless chins, signaled compliance and submission. For the very real Public Health™ purpose of *unquestioning obedience,* masks work very well indeed.

Masks are effective at instilling fear in people. Fearful people more readily submit to authority, particularly when that authority promises a solution to the cause of their fear.

Masks are effective as virtue signals of compliance, bolstering the submissive person's ego.

Masks also impose a very strong peer-pressure effect, which pushes uncertain persons toward following the crowd.

Masks are effective at humiliating people. They are uncomfortable, ugly, dirty, and unnatural. They truly are "face diapers." In a word, masks are *degrading*. If the ways of the old Eastern Bloc taught us anything, it is that the systematic degradation of individuals, especially for patently stupid reasons, is highly effective at promoting totalitarian ends.

Masks are also extremely effective in exposing dissidents. Who dares to stand up against the state? There's one, right over there. Shame on them. Shun them. Arrest them.

That's how masks really "work" and that's why the Public Health™ types love them.

That's why they're trying to bring them back.

My Golden Retriever Confronts the Medical Juggernaut

Originally published October 27, 2023 in Brownstone Journal.

Recently, our golden retriever, Bailey, got kennel cough. She hasn't been in a kennel in years, but that's what they called it: kennel cough.

Please forgive my ignorance in the matter. You see, I'm just a people-doctor. I'm not a veterinarian like, say, Pfizer CEO Albert Bourla. I can't claim to be an expert on kennel cough.

But as far as I can tell, "kennel cough" appears to be vet-speak for a nonspecific respiratory tract infection in dogs. It seems to be a term veterinarians use much as I would "bronchitis."

Do you know what a golden retriever with kennel cough sounds like? After all, people-doctors have historically described kids diagnosed with croup as having a "barking" cough.

Well, based on my limited experience, a golden retriever with kennel cough sounds like a Canada goose. Bailey was repeatedly emitting a medium-pitched grunt/honk, lower in register than a duck's *quack* but higher than one of those old-fashioned *ah-oo-ga* automobile horns.

It's kind of a *Honk! Honk! Honk!* with the H's partially dropped. It's actually quite alarming. Trust me, you don't want to hear your golden retriever sounding like something it retrieved.

Now, Bailey is a good girl, and I love her dearly. But my wife loves that dog more than life itself. Sometimes I wonder if she'd donate her own liver if it were necessary to save her.

So my wife calls Bailey's veterinarian, and she tells them about her symptoms.

I should mention that my wife is a doctor, too. Just a people-doctor like me, mind you, not an expert on kennel cough like Albert Bourla. But a medical case presentation is a medical case presentation, and she knows how to present a case.

So what did Bailey's Primary Care Provider tell my wife after hearing the medical history from a fellow medical professional? Well, they told her that it sounds like kennel cough, and that they can see Bailey in 2 or 3 weeks.

Incidentally, this veterinary practice – I am not making this up – had recently been bought out by some kind of veterinary investment firm which, over the past couple of years, also bought multiple other practices in the area, including the only veterinary emergency room in town. Soon after those acquisitions, they closed down the emergency room.

My wife says to them, "Two or 3 weeks? Bailey will either be fully recovered or dead by then."

"Well, we've been chronically short-staffed," they replied. "We're blocked up for urgent appointments...etc., etc."

A brief, polite back-and-forth ensued, but ultimately Bailey's "provider" didn't offer an urgent appointment.

In their defense, this veterinary group knows what really is important. A couple of months earlier, at Bailey's routine checkup, her doctor noted concerning "plaque buildup" on her teeth.

Do you know what Bailey's doctor recommended? Doggie dental cleaning. Under *general anesthesia*. Seven hundred dollars, cash on the barrelhead.

They also have never delayed care when it comes to Bailey's vaccines.

You see, according to the American Animal Hospital Association Guidelines (generously supported by Boehringer Ingelheim Animal Health, Elanco Animal Health, Merck Animal Health and Zoetis Petcare), all dogs should be vaccinated for:

- Distemper
- Adenovirus
- Parvovirus
- Parainfluenza
- Rabies

while many or most dogs, depending on "lifestyle and risk," should be vaccinated for

- Leptospirosis
- Lyme disease
- *Bordetella*
- Canine influenza

and some should even be inoculated with Rattlesnake Toxoid.

I will add, these vaccines are not one-and-done shots. Most of them are recommended to be boosted annually, or at minimum every 3 years.

But again, the experts know what is really important. For example, while Bailey has fortunately avoided any major orthopedic problems to date, we know at least one golden retriever who has had *both* ACLs reconstructed, and other dogs who have had total hip replacements. Advanced orthopedic surgeries, while admittedly costly, are an essential component of the golden retriever's healthcare armamentarium.

(This probably sounds selfish, but I just hope and pray Bailey doesn't develop gender dysphoria. I don't think we can afford to take her down to Cornell to have them surgically construct a neophallus for her.)

Whew. Let's step back and review. As I said, I'm no expert on these matters, like Albert Bourla. I want to make sure I've got all this correct.

Our golden retriever must navigate a healthcare system that cares so much for her health and well-being that it's willing to intubate and anesthetize her for a tooth cleaning. *Cha-ching!*

In the name of vaccination, it will repeatedly inject her with numerous inoculations, up to and potentially including rattlesnake toxoid. *Cha-ching!*

It offers any number of extensive and expensive Orthopedic surgeries – as long as Bailey's owner pays. *Cha-ching!*

And yet, when she gets sick with an acute respiratory infection, it tells her to stay home and wait, offers no treatment, and refuses to see her. Even though, should she become severely ill, her emergency health care system has been decimated by corporate profiteers.

Do I paint an accurate picture, or do I exaggerate?

Fortunately, Bailey's story has a happy ending.

As so many other concerned patients and family members do, we consulted Dr. Internet. I know, I know, patients are supposed to trust the experts, and refrain from doing their own research – but you'll have to forgive us. After all, it's the *family dog* we're talking about here. And we did discover some interesting information.

According to our research, the most common first-line treatment for kennel cough is doxycycline, an inexpensive, generic, people-antibiotic that's been around since the 1960s. The primary purpose of prescribing it here is to treat against *Bordetella*, the most common bacterial cause of the disease.

Incidentally, Bailey is up-to-date on all her recommended vaccines, so the fact that she got kennel cough in the first place raises its own set of questions. I won't head down that rabbit hole here, except to ask:

If a disease doesn't merit the patient being seen, assessed, and treated when they contract it, why is obsessive vaccination against it so necessary?

My wife called back, and in her very polite but insistent way, explained that if they weren't going to see Bailey, we were 'requesting' a prescription, which in the end they wrote. I half-expected them to say, "Doxycycline, but that's human paste!" To their credit, they didn't.

You'll be glad to hear that after commencing empirical, early treatment with a cheap, decades-old, repurposed drug, Bailey improved almost immediately. Whether this was due to the doxycycline, her own immune system (God gave her one too, we must not forget), or both, we cannot be certain. Anyway, the goose honk is gone, her appetite is back, and she's got the frequent zoomies again.

But the whole episode left *me* with a lingering, uneasy, even unhealthy feeling. It's not exactly déjà vu, but rather the sensation that I'd been through something very similar – and similarly unpleasant – before.

Whatever could that be?

The Depopulation Bomb: A Halloween Sci-Fi Tale

Originally published October 31, 2023
in *Brownstone Journal*.

The following fictional story may or may not bear resemblance to events in real life.

Imagine, if you will, that you are a first-generation high tech gazillionaire. In fact, at one time you were said to be the richest man on earth, although that is no longer the case. Nevertheless, you remain unimaginably wealthy, with all the responsibilities and burdens that such wealth brings. (Given the extremely unusual circumstances of this tale, to make it more relatable, we will assign you a fictional name.) Your birth certificate reads Gilbert Harvey Bates III, but the world knows you as Gil Bates.

Gil Bates's erstwhile net-worth preeminence (stolen as it was by an upstart online retailer named Biff Jezos) is not the only important loss he has suffered. Also in the rearview mirror is his youth, his marriage, and his position as CEO of the behemoth tech company he created, MacroHard™.

After Gil Bates stepped down as CEO of MacroHard™, he focused on his philanthropic work. The centerpiece of this work is the immensely well-funded (and therefore immensely influential) Bates Foundation. The Foundation's scope may be mind-bogglingly broad, but one problem especially consumed Bates: *there are far too many people on the planet.*

In his youth, Gil Bates read a controversial book called *The Overpopulation Bomb*, written by a visionary scientist named Saul Derelicht. That alarming book, a huge bestseller in its day, described a neo-Malthusian hell on earth resulting from human overpopulation, and proposed mass sterilization and other aggressive population reduction techniques as the solution.

Gil Bates became convinced, and remains convinced – especially as the worldwide human population has soared beyond 8 billion units – that *Homo sapiens* have obscenely overpopulated the planet. Once Bates had sold software packages to the great majority of them, he vowed that this existential threat to the planet must be addressed.

But what was to be done? How could this great affront to Gaia be reconciled? When it comes to a responsibility so great, a task so immense, no single man – not even Gil Bates – could hope to accomplish it alone.

Fortunately for the future of Earth, Bates knew a host of like-minded, enlightened elites, pre-eminent individuals of great wealth, power, and worldwide influence. Among the most important:

- A dour Teutonic economist named *Kraut Schlob*. The son of an ambitious industrialist who built flamethrowers for the Third Reich, Schlob is the founder and chairman of the World Enslavement Forum. The Forum has become the premier worldwide gathering of hyper-elites who wish to discuss globalist policies, and enjoy the company of high-end prostitutes, free from the prying eyes of commoners.

- An immensely powerful – if embarrassingly vertically challenged – American health bureaucrat named *Dr. Fantoni Auci*. For decades, Dr. Auci controlled the overwhelming majority of US Government medical research funding. As such, no one in the vast American network of hospitals, research institutes, or universities dares to cross Dr. Auci, and he wields similar influence internationally. In fact, he oversees funding for multiple secret virology research laboratories, as far away as China.
- A mysterious veterinarian named *Adalbert Ghoula*. Ghoula is the CEO of Kaiser, Inc., the world's largest and most rapacious pharmaceutical company, which Ghoula has grown into a veritable modern day IG Farben. In his earlier days, Ghoula oversaw the development of a vaccine that successfully induces the chemical castration and sterilization of swine.

The consensus, reached after lengthy consultations with these men and other luminaries, was that the worldwide human population must be reduced from 8 billion to 500 million units.

But how? Several possible avenues were proposed.

- *War* has been used for millennia to reduce populations, and while highly effective locally or regionally, it would be entirely ineffective at removing the necessary fifteen-sixteenths of people on Earth. After all, the deadliest war in history, World War II, resulted in a mere 80 million

deaths, just 3 percent of the world's population at the time.
- The use of a *bomb* was considered a special kind of bomb, reminiscent of the "neutron bomb" of yore, which would supposedly reduce populations while sparing infrastructure. This seemed closer to the mark than all-out war, but ultimately it was determined that setting off bombs would be both impractical and far too obvious. After all, even herd animals will not consent to being openly and massively slaughtered, no matter how necessary the culling may be. The herd must be kept forever in the dark.
- A plague, a pestilence, a *pandemic* seemed more promising. Past naturally occurring pandemics had reduced human populations much more successfully than wars. The Black Death of 1346-53 may have reduced the world population by as much as 25 percent, a much more encouraging number than the measly 3 percent from World War II. As an added economic bonus, the Black Death served as a very effective concentrator of wealth for the survivors, as it caused minimal collateral property loss.

However, a more detailed review of historical worldwide population estimates demonstrated that a pandemic alone could only serve as a temporizing measure at best. Most estimates show that by 1400, the worldwide population had unfortunately returned to its pre-plague total.

Clearly, the necessary 94 percent reduction in population could not be achieved by culling the herd alone. Sterilization would be needed as well. But how to achieve such mass sterilization?

Many *H. sapiens* possess an intense desire to procreate – that's the source of the problem, after all. Unfortunately, prior historical initiatives for mandatory sterilization – even those of limited scale and scope, such as those targeting the mentally deficient – have met great opposition, at least in the so-called "free" nations.

However, a *vaccine* could be used for mass sterilization. Ghoula's earlier work at Kaiser was proof of this. But a fundamental problem remained: how to get the unsuspecting population – specifically, its children and young adults – to take the stealth-sterilizing inoculation?

The solution, when it came, was a thing of beauty, sublimely subtle and symmetrical. The answer was a two-step process: a pandemic *and* a vaccine. One population reduction device would be released, presented as a worldwide plague. It would be followed by a second population reduction device, presented as the cure.

And the technology was already in place to make it happen. It merely had to be perfected, then enacted.

Employing the Black Magic of gain-of-function virology research, an animal respiratory virus, previously never infecting humans, was genetically engineered to readily infect and spread amongst humans. At a key moment in political history, when a particularly bothersome populist American President named T. Ronald Dump was running for reelection, the virus was released from a Chinese laboratory into the human population.

As the new virus spread, reports of the death and devastation it wrought were spread as well. In actuality, the virus had been engineered so that it was deadly only to the frail, chronically ill, and very old. It was cleverly propagandized, however, as a threat to persons of all ages, a modern day Black Death of sorts.

The US deep state, desperate to disrupt the Dump presidency and remove him from office, were willing partners to manage the control and manipulation of the population through propaganda, and to enforce unprecedented, prolonged lockdowns of society.

Remarkably, they even convinced President Dump to sanction the lockdowns, and to fund the development of the vaccine. Most other countries followed suit.

The new virus rapidly killed off many of the oldest and sickest members of society, as would be expected of a novel respiratory virus. However, the locked-down and isolated populations were barraged with media messages that stirred up mass terror of the virus. Businesses were closed, save for those deemed "essential." Schools were closed, though children were already known to be at statistically zero risk of death. Dissenters were harassed, scapegoated, and punished.

Then, a solution to the pandemic was presented: the vaccine. The vaccine was the savior, the only way out of this crisis.

A few irritating, contrarian dissenters fought back. They protested for civil rights. They stressed the near impossibility of producing an effective vaccine against a rapidly mutating respiratory virus. They identified numerous "safety signals" found in the vaccine trials, and tried to expose these as best they could. But the mainstream media drowned them out, the social media companies (controlled by the deep state) censored them ruthlessly, and after all, once the vaccines were *mandated*, most people took at least a couple of doses.

And the joke was on the dissenters in another, more important respect. These meddlesome do-gooders were indeed intelligent enough to identify the toxicities inherent in the vaccines. But they decried them as "safety signals." The fatal toxicities they identified still seemed to them to be flaws, mistakes, and the unfortunate results of a hasty and mad rush to make money off of the pandemic.

Imagine the naivete.

Early in the vaccine "rollout," young women reported abnormal vaginal bleeding and other menstrual problems after receiving the vaccines, raising concerns about potential unintended consequences to female reproduction. Pathologists found ovaries infiltrated with multiple toxins from the vaccines, both the dreaded "spoke" protein of the virus and "lucid nanoparticles" from the vaccine's delivery system. Even occluded Fallopian tubes were identified.

Soon thereafter, reports appeared in the alternative media of dramatically increased numbers of sudden deaths, primarily in young men, after receiving the vaccine. It often visibly occurred in athletes while on the playing field. This caused considerable alarm, impossible as it was to hide.

In a masterful demonstration of the "limited hangout," officials acknowledged the sudden death phenomenon, but would not even allow mention of the vaccine as a possible cause within the mainstream medical community. Instead, protocols and clinics for this sudden epidemic of heart disease in the young were established, but strangely without any official curiosity as to the cause.

All they knew for sure was that it *couldn't* be the vaccine.

Of course, the infamous "spoke" protein, the same viral antigen chosen by the vaccine's designers to induce the vaccinated patient's body to produce in quantity, just happens to be the most toxic part of the virus. The "spoke" protein deposits itself in tissues throughout the body, wreaking havoc wherever it goes. It has a particular affinity for the heart muscle, causing the inflammatory process known as myocarditis that leads to cardiac arrests.

"Spoke" doesn't stop with the heart, however. It is a remarkably versatile toxin, a sort of Swiss Army monkey wrench in the human body. It causes gigantic, gruesome, rubbery blood clots in the vasculature, seizures in the central nervous system, the aforementioned deposits in ovaries and Fallopian tubes (and testes, for that matter), etcetera, etcetera. What a stroke of genius to choose "spoke" as the antigen the vaccines induce replication of!

The vaccines held another nasty little secret, which even the pathetic, naïve resistance only recognized much later. The vaccines were "contaminated" with plasmids containing MV-40 and MV-40-like DNA sequences. Yes, *that* MV-40, the monkey virus known to cause cancer in multiple animal species.

Could the appearance of so-called "turbo cancers" in vaccinated persons somehow be related to this "contamination?" Well, another limited hangout, this time courtesy of Healthcare Canada, took care of that.

Excess death rates rose dramatically after the vaccine rollout. Birth rates plummeted. To the do-gooders, refuseniks, and dissidents, this was a scandal.

But what did they know? To use a phrase all-too-familiar to the seasoned software developer, these toxicities were not bugs, but features. The vaccines were working exactly as they were supposed to work.

Silly plebes! The "vaccines" were actually a deliberate, multi-pronged, population reduction device. They were *designed* to kill a percentage of young people – mostly male – outright, to poison and disable the female reproductive system at multiple points, and to insert teratogenic plasmids into recipients' cells, to pick off others at undisclosed, later dates. They were merely *packaged and marketed* as a vaccine against a (lab-manufactured) flu-like illness.

As successful as they have been, there remains so much more work yet to be done.

A definite lull occurred in the population's acceptance of repeated injections of the vaccine. The dissidents may be naïve, but they are persistent, and sometimes effective to a degree. But ultimately they will fail.

The general population is lazy, uneducated, and easily terrified. (Some say they are being done a favor by being culled.) They are accustomed to the precedents set by other vaccines. Their reluctance will be worn down with time. Of course respiratory viruses are imperfect targets for vaccines. Once again, that's not a bug, it's a feature! It only means that a new booster of the vaccine will be needed every year – at least.

With each new round of boosters, a new population of girls and young women will be rendered infertile. A new group of boys and young men will suffer cardiac arrest – a very quick and painless way to die, really.

Countless others will contract cancers – *turbo cancer*, to use the current term for these rapidly progressing and deadly malignancies, often of unusual types – bone cancers, muscle cancers, and other former rarities. Not an easy way to die, admittedly. But these tumors mercifully progress to end stage very swiftly, and their value as a population reduction device is undeniable.

Have no fear. It is only a matter of time; only a matter of lather, rinse, repeat. As long as the herd allows itself to be sent through the sheep dip whenever and however often the shepherds proclaim is necessary, *H. sapiens* will get to 500 million. All courtesy of a type of bomb after all, but in this case a microscopic bomb that is released in each person via a tiny little injection: *The Depopulation Bomb.*

Happy Halloween!

What is Medical Freedom, Exactly?

Originally Published November 20, 2023
in *Brownstone Journal.*

The beginning of wisdom is the definition of terms - **SOCRATES**

The phrase "medical freedom" has become common usage in the wake of the Covid-19 catastrophe. But like many buzzwords and neologisms, "medical freedom" is perhaps ill-defined or even undefined. We all know more or less what it means in our own minds, or at least we think we do. But when speaking about medical freedom with others, are we talking about the same thing?

In fact, "medical freedom" has become more than a buzzword. It is also a movement, with its advocates, experts, and critics. Multiple medical freedom conferences have been organized and are taking place in the United States and abroad, and political parties under its banner have formed.

As Socrates warns, the lack of a standard definition for an important concept, much less an active movement, is a problem. Like the proverbial blind men describing an elephant to each other, when we lack a standard definition, persons with different perspectives end up talking at rather than to each other about different ideas, while thinking they are meaningfully communicating about the same thing.

What follows is a brief summary of my efforts to find a standard definition of medical freedom. (Spoiler alert: I failed to find one, so I wrote the best definition that I could.)

For what it's worth, *Wikipedia* does not have an entry for "medical freedom" as of this writing. However, it defines "health freedom" as follows: "The health freedom movement is a libertarian coalition that opposes regulation of health practices, and advocates for increased access to "non-traditional" health care."

It goes on to associate said movement with such luminaries as former Congressman Ron Paul, former Beatle Paul McCartney, and yes, the John Birch Society.

In the mainstream media, starting about 2 years ago – soon after the onset of the Covid-19 vaccine mandates – published articles appeared that characterized "medical freedom," at least in part, as a sort of rallying cry for right-wing militia initiatives.

For example, in an article dated August 7, 2021, the *Washington Post* reported on the then-burgeoning medical freedom movement in Western New York. The *Post* described the movement as a recruitment tool for far-right militia groups, even referencing the remote and entirely unrelated incidents of Ruby Ridge, Idaho, Waco, Texas, and even the Oklahoma City bombing. The *Post* article states:

Far-right groups have aligned themselves with those opposed to masks and vaccines, seeking new allies around the issue of "medical freedom" while appearing to downplay their traditional focus on guns, belief in the tyranny of the federal government and calls by some for violent resistance.

Notably, the article's author, one Razzan Nakhlawi, is currently listed on the *Post* website as "a researcher on the *Post*'s National Security desk."

More recently, with public distrust in vaccines reaching historic highs, the media has shifted its characterization of medical freedom from a domestic terror threat to a cabal of ingenious and industrious hucksters. (After all, how can a few crackpot far-right militiamen sway mass public opinion so successfully?)

In a March 24, 2023 article, the far-left magazine the *Nation* described "The Medical Freedom Hustle" as follows:

> Under the great dispensation of our new age of medical freedom, these disparate forces—ambitious Republican politicians, self-interested medical professionals, profiteering quacks, and nihilist visionaries—have melded.

It would be a subject for another day and another essay to unpack all the psychological projection concentrated in that quote. Suffice it to say that the traditional far left – insofar as outlets like *the Nation* represent it – has come to characterize "medical freedom" largely as a kind of scam or confidence game, allegedly designed to draw the population away from legitimate mainstream medicine and toward the folly of snake-oil and naturopathic quackery.

Those more supportive of "medical freedom" see it very differently than legacy media such as the *Post* or far-left outlets like *the Nation*.

Florida Governor Ron DeSantis has declared his state "the Medical Freedom State." In May of 2023 he signed 4 pieces of legislation that were touted as "the strongest legislation in the Nation for medical freedom." Most prominent among these was

Senate Bill 252 – Most Comprehensive Medical Freedom Bill in the Nation:

- Prohibiting business and governmental entities from requiring individuals to provide proof of vaccination or post-infection recovery from any disease to gain access to, entry upon, or service from such entities.
- Prohibiting employers from refusing employment to or discharging, disciplining, demoting, or otherwise discriminating against an individual solely on the basis of vaccination or immunity status.
- Preventing discrimination against Floridians related to Covid-19 vaccination or immunity status, etc.

The other 3 laws 1) banned gain-of-function research in Florida, 2) provided protections for physicians' freedom of speech, and 3) provided "an exemption from public records requirements for certain information relating to complaints or investigations regarding violations of provisions protecting from discrimination based on health care choices."

As politics is, in Bismarck's words, "the art of the possible," it is difficult at best to reverse-engineer passed legislation into a clear understanding of the underlying principles that generated it.

However, it does appear that the Florida "medical freedom"

legislation attempts to address aspects of 3 problems that became obvious during the Covid-19 era. These are 1) the medical and public health infringement on citizens' fundamental civil liberties, 2) the systematic and oppressive control and silencing of physicians during the pandemic, and 3) the apparently out-of-control, dangerous, and unethical research that spawned the pandemic in the first place.

Extrapolated further, these pieces of legislation appear to be steps toward reestablishing 3 things: patient autonomy, physician autonomy, and truly ethical practice across all of medicine, from bench research to bedside patient care.

The Medical Freedom Party, a political party formed in New York City in April 2022 in the wake of Covid-19 mandates, states in its platform:

> The Medical Freedom Party believes the individual is endowed by his or her creator with the inalienable right to bodily autonomy. The Medical Freedom Party asserts that bodily autonomy is the basis from which all freedoms flow.

The party's platform goes on to make several more detailed assertions, all of which expand on their insistence for absolute bodily autonomy. This appears to be their principal and perhaps overwhelming concern with regard to medical freedom.

Also notable in their platform is their clear use of language from the Declaration of Independence. To them, bodily autonomy is a fundamental right, fully equivalent to life, liberty, and the pursuit of happiness.

While this points us in a clearer direction regarding the priorities and views of medical freedom advocates, we still lack an explicit definition for medical freedom. Furthermore, it becomes apparent that different groups may focus on one particular part of the concept, possibly ignoring or underestimating the importance of others.

I would like to propose my definition of medical freedom here.

I submit it as a serious and genuine effort at establishing a sound working definition for this important concept, so that interested parties discussing medical freedom can be confident that they are speaking about the same thing. I welcome discussion about its finer points, or even its larger ones, as others feel necessary. After all, that is one of the prime purposes of a working definition – to invite discussion and to work toward the best consensus possible.

In my research, I drew on conversations from many colleagues who are knowledgeable on this issue. I also referred to foundational medical ethics writings, many of which I have written about in the past.

As an American, I also referred in detail to the founding documents of our country, specifically the Declaration of Independence and the Bill of Rights. I did so for a couple of reasons. First, they are commonly cited by medical freedom advocates, as seen above. Second, it is undeniable that in the name of "public health," numerous freedoms clearly stated in the Bill of Rights were taken away from citizens during the Covid-19 lockdowns, by extralegal executive fiat, at multiple levels of government.

Finally, I made a genuine effort to assess negative views of the concept, such as those at the beginning of this essay. Ultimately, I must admit I gave up in cases such as those cited above. I believe that many of these characterizations from the mainstream media and/or the far left have been made in conscious bad faith. I have come to know numerous medical freedom advocates, and accusations, for example, that they are the tools of covert, nascent Timothy McVeighs are too patently absurd not only for me to believe, but for me to believe the purveyors of such claims believe themselves.

One can be opposed to a concept and still be willing to work toward a rational definition of it. I am personally opposed to communism, but I am able to refer to it, at least definitionally, as something like "a Marxist, socialist economic theory whereby Government controls all means of production, in pursuit of a classless society."

If I refuse to accept any definition other than "a bunch of murderous bastards," then there's not much hope discussing its pros and cons, is there? I fear this is more or less where we are, at least at present, with many opponents to the notion of medical freedom.

I sought to make my definition broad enough to cover all of the main ideas it must contain, but brief enough to be useful and memorable. I settled upon a 3-part definition.

One might think of this definition of medical freedom as something like a three-legged stool. All 3 legs must be in place for the stool to remain standing. The first component (or "leg") of medical freedom focuses on the individual patient, the second addresses public health and providers of health care, and the third emphasizes the philosophical, ethical, and even legal underpinnings of the concept.

I addended the definition with a longer list of related but subsidiary concepts that I felt must be considered as well. If one envisions the definition *per se* as a sort of "Declaration of Independence," the list that follows it might be thought of as analogous to a "Bill of Rights."

Here is my definition of medical freedom:

Medical freedom is a moral, ethical, and legal concept, essential to the just and proper practice of medicine, that asserts the following:

1. The individual patient's autonomy over his or her own body with regard to any and all medical treatment is absolute and inalienable.
2. Physicians and public health officials do not possess the authority to deprive any citizen of their fundamental civil rights, including during a declared medical emergency.
3. The four fundamental pillars of medical ethics – autonomy, beneficence, non-maleficence, and justice – are essential to medical practice and must be observed at all times by all physicians, nurses, public health officials, researchers, manufacturers, and all others involved in health care.

In the wake of the Covid-19 catastrophe, and in light of the innumerable abuses and affronts to basic civil rights that the public health establishment and the physicians under them inflicted upon citizens, several derivative statements follow.

1. Patient autonomy depends upon informed consent, confidentiality, truth-telling, and protection against coercion.
2. Informed consent must be obtained for all health care interventions, including but not limited to invasive procedures, vaccinations, and medications. To be valid, informed consent requires a competent patient (or a competent proxy representing the patient's best interest) who receives full disclosure, and after understanding it, voluntarily agrees.
3. Confidentiality is central to patient autonomy. Specifically, any "health passport" type of public health approach violates patient autonomy, and must be forbidden.

4. Truth-telling. Physicians and health officials are duty-bound to tell the truth. Willful deviation from this violates patient autonomy, and must result in professional discipline.
5. Coercion of any kind, applied to patients or health care providers, violates patient autonomy. This includes bribery, incentivization, threats, blackmail, public shaming, scapegoating, exclusion or ostracization from society, deceptive advertising, and all other forms of coercion.
6. Beneficence requires that all treatments given to a patient should be done only when the prospect, intention, and likelihood of providing genuine benefit to that patient exists. There must be no "taking one for the team."
7. Non-maleficence refers to the "First, do no harm" precept of medical practice. No medical treatment should be imposed on any patient that is likely to harm the patient, or where the risk/benefit ratio is negative for that patient.
8. Justice requires that both the benefits and burdens of medical care must be distributed equally throughout the population. A new emphasis on the protection of vulnerable populations, especially children, is essential.
9. Public health directives that impact citizens' civil rights in any way must be enacted lawfully through legislation, not by emergency declaration or by executive or bureaucratic fiat.
10. Refusal of treatment should never result in punishment. Specifically, it must not preclude a patient from receiving other treatments, except where the first treatment is an absolute medical prerequisite for the second treatment.
11. Open and honest debate. The medical profession must

allow, and indeed encourage, open and honest debate within its ranks, without fear of reprisal.

12. Censorship, silencing, intimidation, and punishment of physicians and other health providers for making statements contrary to the officially approved or majority medical narrative must be prohibited, under penalty of professional and/or legal punishment of the censors.
13. Patient redress. Patients must have the right to seek real and meaningful redress for any kind of negligent or malicious harm done to them by any physicians, health care systems, public health officials, or producers of drugs or other health care products. No one involved in the healthcare enterprise may be immune, and laws providing such immunity must be removed.
14. Outside influences. The medical profession must eliminate all undue outside influences from its decision-making process, including financial incentives from industry, private foundations, insurance companies, and unelected international entities.
15. The patient-physician partnership. The patient, working one-on-one with their physician, must make clinical care decisions, with the patient reserving ultimate authority to decide. Clinical care decisions must not be predetermined by government bureaucrats, statistical analyses, industry influence, insurance carriers, or other outside influences.
16. Protocols. The mandated or coerced use of strict or inflexible protocols in medical practice must be prohibited. Variation from protocols, to allow for individualized patient care decisions, must be allowed.

Multiple public health officials, including current CDC Director Mandy Cohen, have noted the loss of public trust in the medical establishment, the public health enterprise, and physicians in general, in the wake of Covid-19. While they are correct that trust has been lost, many appear oblivious to the reason for it, namely the appalling abuses of power they themselves oversaw during the Covid-19 era.

The only real way to restore public trust in medicine is for those in charge to acknowledge their wrongdoing, accept responsibility for it, and for medicine to reform, from the oppressive and overbearing population-based system of the Covid-19 era, into a truly patient-centered system that serves the individual patient first and foremost.

I am hopeful that this definition of medical freedom – and the "bill of rights" that follows from it – will invite productive discussion and debate, and will prove beneficial to this vitally important process of reforming the entire medical enterprise.

Acknowledgments: When writing this essay, I drew from conversations and communications with numerous people who are knowledgeable on the subject in question. These include (but are not limited to): Kelly Victory MD, Meryl Nass MD, Kat Lindley MD, Peter McCullough MD, Ahmad Malik MD, Drew Pinsky MD, Jane Orient MD, Lucia Sinatra, Bobbie Anne Cox, Tom Harrington, Shannon Joy, and my editor Jeffrey Tucker. I am gratefully indebted to these people. They deserve recognition for much of whatever is of value here. For any errors, confusion, or dross, I claim full credit.

Ten New Year's Resolutions to Restore Medical Freedom

*Originally published on December 30, 2023
in Brownstone Journal.*

As 2023 staggers to its conclusion, leaving behind a world of brutal wars, tenuous economies, corrupt governments, and tyrannical elites, perhaps the most unsettling aspect of the year's end is a strange silence.

Some things always generate plenty of noise. The 2024 US Presidential election promises to be even more hysterical than the last two. It will probably be a rematch, pitting a widely hated octogenarian incumbent President with obvious, rapidly progressing dementia against a widely hated late-septuagenarian former President facing dozens of felony indictments. Still almost a year away, the commotion surrounding this impending showdown of the senescent is already continuous, cacophonous, and confounding.

However, regarding the most important historical event since World War II, there is almost total silence.

The Covid-19 debacle is the defining event of the 21st century. It is at once the worst act of biological warfare in human history and the greatest mass violation of civil liberties since the Iron Curtain. Even more importantly, it is the self-evident template for the establishment of the technocratic soft-core totalitarianism advocated by globalist entities such as the World Health Organization and the World Economic Forum.

And yet virtually no one in the mainstream will discuss it. The legacy media shows near zero curiosity regarding Covid's origins, the disastrous response, or the toxic vaccines.

Both the Biden and Trump camps pretend it never happened. Out of the 4 Republican debates held to date, only *one* question has been asked about Covid vaccines. And that single exchange, between journalist Megyn Kelly and candidate Vivek Ramaswamy, was mysteriously blacked out, even from supposedly "free speech" platform Rumble's livestream of the event, with Rumble's CEO later blaming the blackout on "the source feed from a 3rd party" which he did not name. Nothing to see here.

Among the other presidential candidates, former Democrat Robert F. Kennedy, Jr. and Republican Ron DeSantis have spoken up repeatedly and honestly about Covid. As a result, they have both been aggressively reviled and ostracized by both the mainstream media and the establishments of both political parties.

Advocates for civil rights in general, and for medical freedom in particular, should be deeply disturbed by this attempt to cast the whole Covid-19 catastrophe down the memory hole. Medical freedom is rapidly developing as a philosophical, intellectual, and ethical concept.

However, theoretical efforts to promote medical freedom – and by extension, to re-enforce all fundamental civil liberties – will come to naught if the greatest assault on freedom in modern history is allowed to be forgotten, and the perpetrators are allowed to continue as if nothing happened.

As a prominent man once asked: "What is to be done?" In my attempt to answer that question, here are **10 New Year's Resolutions for Medical Freedom advocates.**

1. Speak the Truth About Covid at Every Opportunity.

Honest and informed citizens, politicians, and public figures must plainly tell truthful narratives about Covid every chance they get. A brief, factual account might sound something like this:

a. SARS CoV-2 is a man-made bioweapon developed through US Government funding, which got out of the lab and into the human population.
b. The mRNA Covid vaccines are essentially pre-planned antidotes to that bioweapon, which were hastily produced and aggressively pushed on the population for profit, with an appalling and criminal disregard for safety.
c. The lockdowns, masking, school closures, mandates, censorship, scapegoating, etc., were deliberate and illegal assaults on citizens' civil rights – blatant power grabs that governments made under the pretense of a declared emergency.

Medical freedom advocates must explain to people that they have been repeatedly lied to for the past 4 years, by virtually every authority. Then, tell them the truth – coolly, rationally, and politely. If they don't want to hear it, tell them anyway.

For decades, every citizen in modern Western society has been browbeaten with leftist and globalist propaganda, ranging from countless Global Warming false prophecies, to risible DEI nonsense, to Baskin-Robbinsesque gender insanity, to fascistic vaccine absolutism. Then came Covid. At this late date, it is reasonable and salutary to present one's neighbor with a brief smattering of truth.

2. Encourage and Petition Politicians to Commit to Medical Freedom Policies.

The Pharma industry spent a reported $379 million on political lobbying in 2022 alone. It's going to take a lot of grassroots work with politicians to combat the pernicious influence of that much purchased influence.

There is evidence that this can be done. People such as Dr. Mary Talley Bowden in Texas are leading the way in this regard. As of December 23, 2023, Bowden and colleagues have convinced 40 candidates and 25 elected officials from 17 states to publicly state that "the Covid shots must be pulled off the market." Per Dr. Bowden, "many of these are also pledging not to take donations from Big Pharma."

Those committed to medical freedom should set all their elected officials and relevant appointed government bureaucrats on speed dial. These individuals in positions of power – at all levels, local to national – must hear regularly from their constituents. Constituents must tell these people exactly what they *know*, as well as what they want. It is now up to constituents to teach their officials the facts about the world.

As Andrew Lowenthal has demonstrated in detail, the Censorship Industrial Complex is real, and because of it, many elected officials and bureaucrats suffer from the same lack of accurate information on policy matters as the majority of their constituents.

3. Work to Outlaw all Gain-of-Function Research.

All research regarding the genetic manipulation of viruses needs to end. Robert F. Kennedy, Jr. and others have pointed out that such research is really bioweapons research, in which our tax dollars are misused to fund the development of a bioweapon and its antidote vaccine in concert. In Florida, Governor Ron DeSantis and the state legislature have passed laws banning gain-of-function research in that state.

The Covid era displayed in high relief the disastrous wages of such "research." It needs to be completely outlawed *everywhere*, and all labs involved in such work, from the Wuhan Institute of Virology, to the Ralph Baric lab at the University of North Carolina, to illegal labs in the rural US or allegedly in places like the Ukraine, need to be permanently shut down.

Key to achieving this is not falling prey to the intentionally confusing semantic arguments about what technically constitutes "Gain-of-function" and what doesn't. The word games Anthony Fauci played with Congress need to be called out as the dishonest prevarications they are, and rejected as a defense for those involved in such wicked "research." (Of note, the Florida laws included language to prevent this deception, outlawing all "enhanced potential pandemic pathogen research.")

4. Work to Get the US out of the World Health Organization.

The WHO's newly proposed pandemic agreement and amendments to the existing International Health Regulations (IHR) unfortunately are bald-faced, bad-faith attempts to usurp power from sovereign nations by an unelected globalist elite, all in the nebulous name of "global health."

As David Bell and Thi Thuy Van Dinh have written, despite

claims by WHO Director General Tedros Ghebreyesus that "no country will cede any sovereignty to [the] WHO," in fact

1. The documents propose a transfer of decision-making power to the WHO regarding basic aspects of societal function, which countries *undertake* to enact.
2. The WHO Director-General will have sole authority to decide when and where they are applied.
3. The proposals are intended to be binding under international law.

Furthermore, the proposed amendments to the IHR will change WHO directives during declared health emergencies from non-binding recommendations to dictates with the force of international law. As Bell and Dinh state, "It seems outrageous from a human rights perspective that the amendments will enable the WHO to dictate countries to require individual medical examinations and vaccinations whenever it declares a pandemic."

And the potential incursions to medical freedom hardly end there, potentially including all the items in Article 18 of the existing IHR, which already directly contradict the UN's own Universal Declaration of Human Rights in multiple places.

Most current debate on the matter surrounds the question of whether individual countries should accept or reject these proposals. However, in the wake of the Covid disaster, the WHO's current proposals reveal that its intention is not to step back, learn from the catastrophe, and account for the mistakes it and other authorities made. Rather, it seeks to consolidate its own power by permanently encoding the top-down, public-health-by-totalitarian-diktat approach that caused so much destruction. Not only these policies, but the organization proposing them should be categorically rejected.

The WHO is a classic wolf in sheep's clothing. It is an unelected globalist cabal of profiteering elites, heavily funded by Bill Gates and closely associated with the World Economic Forum. It is engaged in blatant political power-grabbing while masquerading as a benevolent public health institution.

It is insufficient for nations to merely reject the WHO's proposed pandemic agreement and amendments to its IHR. The US and every sovereign nation should leave the WHO entirely, and medical freedom advocates should lead the way in the struggle to make this happen.

5. Join the Fight to Remove the Covid mRNA Vaccines From the Market.

The Covid-19 mRNA vaccines have demonstrated toxicities far more common, more varied, and more severe than numerous conventional medicines that have been appropriately pulled from the market in the past. Dr. Peter McCollough and numerous other leaders in the fight for medical freedom have rightly called for the Covid mRNA vaccines to be removed from the market.

Despite the intense efforts of Big Pharma, the growing Censorship Industrial Complex, and captured government agencies, public awareness of the numerous and often deadly toxicities of the Covid mRNA injections is growing.

This is reflected in both reduced public "uptake" for recurrent "boosters" per CDC data and the falling stock price of Pfizer, Inc. A small but growing number of politicians, as described above, are committing to the fight to remove the vaccines from the market, demonstrating that this is becoming a tenable and perhaps winning political position to hold.

Encouraging as these trends may be, they are insufficient

on their own. Medical freedom advocates should speak out supporting the removal of the Covid mRNA vaccines from the market. They should recruit, support, and vote for elected officials and candidates taking this position, and support legal actions toward this goal.

6. Push for a Moratorium on the mRNA-Based Pharmaceutical Platform as a Whole.

Even if the Covid mRNA vaccines are removed from the market, a widely overlooked corollary question remains: how much of the toxicity from these products is Covid-specific, i.e. due to the spike protein, and how much is due to the deeply problematic and incompletely understood mRNA platform itself?

There is certainly plenty of toxicity to go around, as numerous mechanisms of injury have been identified from these injections. These include toxicities to the heart, immune system, skin, reproductive organs, blood clotting cascade, and cancer promotion, among others. It is willful denial at best and criminal negligence at worst to assume that the mRNA platform does not contribute to these problems.

mRNA vaccines are currently in use in food animals, notably swine. Furthermore on its own website, Moderna describes a pipeline of mRNA vaccines currently in development for Influenza, Respiratory Syncytial Virus (RSV), Cytomegalovirus (CMV), Epstein-Barr Virus (EBV), Human Immunodeficiency Virus (HIV), Norovirus, Lyme disease, Zika virus, Nipah virus, Monkeypox, and others. Meanwhile, the trial for its EBV vaccine has reportedly been halted in adolescents due to a case of – you guessed it – myocarditis.

The human population will soon be inundated with mRNA-based drugs on a scale and with an imposed intensity that will make the Covid era seem positively quaint. The safety record for the sole mRNA product currently in human use – the Covid vaccines – is abysmal.

A moratorium of at least several years, combined with an open, thorough, and publicly debated inquiry into the likely and possible toxicities inherent to the mRNA platform is essential to human safety, and if done, will save countless lives in coming years.

7. Work to Have the 1986 Vaccine Act Repealed.

The toxicity of vaccines was so well-established even decades ago, that a Federal law – the National Childhood Vaccine Injury Act (NCVIA) of 1986 (42 U.S.C. §§ 300aa-1 to 300aa-34) – was passed to specifically exempt vaccine manufacturers from product liability, based on the legal principle that vaccines are "unavoidably unsafe" products.

Since the 1986 NCVIA act protecting vaccine manufacturers from liability, there has been a dramatic increase in the number of vaccines on the market, as well as the number of vaccines added to the CDC vaccine schedules, with the number of vaccines on the CDC Child and Adolescent schedule rising from 7 in 1986 to 21 in 2023.

The National Childhood Vaccine Injury Act (NCVIA) of 1986 should be repealed, returning vaccines to the same liability status as other drugs.

8. Work to End Vaccine Mandates at Every Level of Society.

According to the National Center for Education Statistics, in the 2019-20 academic year there were 3,982 degree-granting colleges and universities in the United States. In the fall of 2021, all but approximately 600 of these institutions mandated Covid-19 vaccination for their students.

Since then, nearly all such institutions have dropped their

student Covid vaccine mandates. However, at this writing, 71 colleges and universities, or approximately 1.7%, continue to mandate the Covid vaccines for students to attend.

The number of mandating schools reduced gradually, largely through the intense, extremely labor intensive work of a very few small, newly-formed, grassroots organizations like No College Mandates. While the effectiveness of such efforts is undeniable, the 71 holdouts (which include "elite" institutions such as Harvard and Johns Hopkins) demonstrate just how deeply entrenched the mandating of vaccines remains in certain segments of society.

As a result of the hubris and abuse it displayed during Covid, the entire vaccine mega-industry has suffered tremendous (and richly deserved) damage to its formerly unquestioned, "safe and effective" image. However, from education to healthcare to the military, gains made against vaccine mandates have been partial and temporary at most. A concerted effort to further educate the public about the immense problems with vaccines and to restore individual choice must be joined by a great many more people if this fundamental imposition on basic bodily autonomy is to be overcome.

9. Work to End Direct-to-Consumer Advertising of Pharmaceuticals.

The United States is one of only 2 countries in the world that allows direct-to-consumer advertising of pharmaceuticals. The dangers of this utterly ill-advised policy are multiple.

First, as we all can see by simply turning on the television, Big Pharma abuses this privilege to aggressively yet seductively hawk every product it feels it can make a buck off of. The "pill for every ill" mindset shifts into hyperdrive, with an expensive, proprietary, pharmacological cure for everything from your morbid obesity to your "bent carrot." The situation on social media is, if anything, even worse.

It is no coincidence that black markets for overhyped, purported wonder drugs such as semaglutide develop, nor that dangerous misuse, such as thousands of reported overdoses have been reported. Perhaps more importantly, direct-to-consumer advertising provides Big Pharma with a convenient and legal way to capture media. Big Pharma was the second-largest television advertising industry in 2021, spending $5.6 billion. No legacy media outlet dares to go against the wishes of those providing that level of funding. This effectively muzzles any and all dissenting voices from appearing on those platforms.

A free society requires freedom of the press and media. The Covid era has demonstrated that direct-to-consumer pharmaceutical advertising stifles freedom of the press and media to a dangerous and unacceptable degree.

10. Play Offense.

If all you do is play defense, the best result you can hope for is a draw. During the lockdowns, with courts closed, businesses shuttered, and citizens isolated from one another, it was extremely difficult to mount even a solid defense against the gross incursions on our civil rights. A few courageous individuals, often acting alone and at tremendous personal cost, managed to counterpunch effectively. Their contributions to saving our "free" societies (if indeed they are eventually saved) will perhaps never be adequately recognized.

Today, despite the mainstream silence, the tide is turning in favor of medical freedom and civil liberties in multiple areas. It is time for the masses to join in and help those who managed to make these early advances, and who continue to fight on behalf of all citizens.

For example, New York attorney Bobbie Anne Cox continues her David v. Goliath legal struggle to defeat Governor Kathy

Hochul's extralegal and grossly unconstitutional quarantine camp order. This case may eventually reach the Supreme Court. I don't want to declare that Ms. Cox can't do it alone, because that's pretty much what she has done so far, and having followed that case, I wouldn't bet against her. But hell, even Hercules had a sidekick. Medical freedom advocates would do well by actively and generously supporting her.

After surviving his own trial by fire, Texas Attorney General Ken Paxton has announced a lawsuit against Pfizer for "unlawfully misrepresenting the Covid-19 vaccine's effectiveness, and attempting to censor public discussion of the product." Citizens of other states would do well to aggressively petition their attorneys general to take similar action, including removing the mRNA vaccines from the market in their states on the grounds of their demonstrated adulteration with potentially harmful DNA.

If medical freedom advocates want the concept to prevail, they must go on offense. Get involved. There is no need to reinvent the wheel at this point. Adopt one or more of the organizations or causes above as your personal project, join, and contribute. Add your light to the sum of light, and the darkness will not overcome it.

In summary, those of us seeking to secure and ensure medical freedom for ourselves and future generations must become vocal, persistent advocates, as well as courageous people of action.

Furthermore, we must not allow the abuses and evils of the Covid era to vanish down the memory hole, which of course is exactly what every politician, bureaucrat, Deep State apparatchik, and globalist elite who perpetrated those deeds wants to happen. Some cliches are true, and this is one of them: if we allow ourselves to forget history, we will be doomed to repeat it.

Covid-19 was the defining event of the century. It was a destructive, deadly catastrophe, but it does have one remarkable silver lining. It peeled the veneer off our governments, institutions, corporations, and society as a whole. It revealed how the powerful plan to strip us of our freedoms – medical and otherwise. We now know what we face. May we, the ordinary citizens, have the courage and intelligence to act effectively to regain and retain our freedoms, dignity, and fundamental human rights.

Medicine Has Been Fully Militarized

Originally published January 30, 2024 in *Brownstone Journal*, January 31, 2024 in *RealClearHealth*, and February 6, 2024 *in Epoch Times*.

I am thinking of a certain industry. See if you can guess what it is.

This industry is huge, constituting a large portion of the nation's GDP. Millions of people earn their living through it, directly or indirectly. The people at the top of this industry (who operate mostly behind the scenes, of course) are among the super-rich. This industry's corporations lobby the nation's government relentlessly, to the tune of billions of dollars per year, both to secure lucrative contracts and to influence national policy in their favor. This investment pays off richly, sometimes reaching trillions of dollars.

The corporations supplying this industry with its materiel conduct advanced, highly technical research that is far beyond the understanding of the average citizen. The citizens fund this research, however, through tax dollars. Unbeknownst to them, many of the profits gained from the products developed using tax dollars are kept by the corporations' executives and investors.

This industry addresses fundamental, life-or-death issues facing the nation. As such, it relentlessly promotes itself as a global force for good, claiming to protect and save countless lives. However, it kills a lot of people too, and the balance is not always a favorable one.

The operational side of this industry is emphatically top-down in its structure and function. Those who work at the ground level must undergo rigorous training that standardizes their attitudes and behavior. They must follow strict codes of practice, and they are subject to harsh professional discipline if they deviate from accepted policies and procedures, or even if they publicly question them.

Finally, these ground-level personnel are handled in a peculiar manner. Publicly, they are frequently lauded as heroes, particularly under declared periods of crisis. Privately, they are kept completely in the dark regarding high-level industry decisions, and they are often lied to outright by those at higher levels of command. The "grunts" even significantly forfeit some fundamental civil liberties for the privilege of working in the industry.

What industry am I describing?

If you answered, "the military," of course you would be correct. However, if you answered "the medical industry," you would be every bit as right.

In President Eisenhower's farewell speech of January 17, 1961, he stated that "…in the councils of government, we must guard against the acquisition of unwarranted influence, whether sought or unsought, by the military-industrial complex." Sixty-three years on, many Americans understand what he was referring to.

They see the endless cycle of undeclared wars and decades-long foreign occupations that are undertaken on nebulous or even outright false pretenses. They see the ever-hungry mega-industry that produces super-expensive, high-tech killing devices of every imaginable form, as well as the steady stream of traumatized soldiers that it spits out. War (or, if you prefer its Orwellian nickname, "defense") is big business. And as Eisenhower warned, as long as those profiting from it drive the policy and the money stream, it will not only continue, it will continue to grow.

Other mega-industries – the medical industry in particular – have generally fared better in public perception than the military-industrial complex. Then came Covid.

Among its many harsh lessons, Covid has taught us this: if you substitute Pfizer and Moderna for Raytheon and Lockheed Martin, and swap the NIH and CDC for the Pentagon, you get the same result. The "medical-industrial complex" is every bit as real as its military-industrial counterpart, and it is every bit as real a problem.

As a physician, I am embarrassed to admit that until Covid, I possessed only an inkling that this was so – or more accurately, I knew it, but didn't realize how bad it was, and I didn't worry about it too much. Sure (I thought), Pharma engaged in dishonest practices, but we'd known that for decades, and after all, they do make some effective drugs. Yes, physicians were increasingly becoming employees, and protocols were dictating care more and more, but the profession still seemed manageable. True, healthcare was far too expensive (gobbling up a reported 18.3 percent of the US GDP in 2021), but healthcare is inherently expensive. And after all, we were saving lives.

Until we weren't.

By early-to-mid 2020, it became obvious to those paying attention that the Covid "response," while promoted as a medical initiative, was in fact a military operation. Martial law had effectively been declared approximately on the Ides of March 2020, after President Trump was mysteriously convinced to cede the Covid response (and practically speaking, control of the nation) to the National Security Council. Civil liberties – freedom of assembly, worship, the right to travel, to earn one's living, to pursue one's education, to obtain legal relief – were rendered null and void.

Top-down diktats on how to manage Covid patients were handed down to physicians from high above, and these were enforced with a militaristic rigidity unseen in doctors' professional lifetimes. The mandated protocols made no sense. They ignored fundamental tenets of both sound medical practice and medical ethics. They shamelessly lied about well-known, tried-and-true medicines that were known to be safe and appeared to work. The protocols killed people.

Those physicians and other professionals who spoke out were effectively court-martialed. State medical boards, specialty certification boards, and large healthcare system employers virtually tripped over each other in the rush to delicense, decertify, and fire dissenters. Genuine, courageous physicians who actually treat patients, such as Peter McCullough, Mary Talley Bowden, Scott Jensen, Simone Gold, and others, were persecuted, while non-practicing bureaucrats like Anthony Fauci were hailed with false titles like "America's Top Doctor." The propaganda was as nauseating as it was blatant. And then came the jabs.

How did this happen to medicine?

It all seemed so sudden, but in fact it has been in the works for years.

Covid taught us (by the way, Covid has been such a harsh tutor, but haven't we learned *so much* from her!) that the medical-industrial complex and military-industrial complex are deeply connected. They are not just twins, or even identical twins. They are *conjoined* twins, and so-called "Public Health" is the tissue shared between them.

The SARS CoV-2 virus, after all, is a bioweapon, developed over a period of years, funded by US tax dollars in a joint effort between Fauci's NIH and the Department of Defense to genetically manipulate the transmissibility and virulence of coronaviruses (all done in the name of "Public Health," of course).

Once the bioweapon was out of the lab and into the population, the race was on within the medical-industrial complex to develop and market the supremely profitable antidote to the bioweapon. Cue the full-on military takeover of medicine: the martial law lockdowns, the suppression of cheap and effective treatments, the persecution of dissidents, the ceaseless propaganda and anti-science, and the unabashed whoring of most hospital systems for CARES Act money.

We know the rest. The ill-conceived, toxic, gene-therapy antidote, falsely billed as a "vaccine," was foisted upon the population by blackmail ("the vaccine is how we end the pandemic"), the effective bribery of medical authorities and politicians, as well as other Deep-State directed psyops designed to divide the population and scapegoat dissenters ("pandemic of the unvaccinated").

The end result even sounds like the aftermath of a gigantic military operation. Millions are dead, many millions more are psychologically traumatized, economies are in tatters, and a few warmongers are fantastically rich. Moderna CEO Stephane Bancel (who, incidentally, oversaw the construction of the Wuhan Institute of Virology years ago) is a freshly minted billionaire.

And not one of those who caused all the mischief are in prison.

At this writing, virtually all the major healthcare systems, specialty regulatory boards, specialty associations, and medical schools are standing at attention, still in lockstep with the received – and by now, clearly false – narrative. Their funding, after all, be it from Pharma or the Government, depends upon their obedience. Barring dramatic change, they will respond in the same fashion when orders come down from above in the future. Medicine has been fully militarized.

In his farewell address, Eisenhower said something else that I believe is most prescient here. He described that a military-industrial complex fostered "a recurring temptation to feel that some spectacular and costly action could become the miraculous solution to all current difficulties."

Enter Disease X.

France Teeters on the Brink

Originally published Febraury 20, 2024
in *Brownstone Journal*.

In viewing this tragi-comic scene,
the most opposite passions necessarily succeed,
and sometimes mix with each other in the mind;
alternate contempt and indignation;
alternate laughter and tears;
alternate scorn and horror.

EDMUND BURKE

Upon first seeing videos of the European farmers' recent protests, I, along with many others across the Atlantic, was deeply impressed. Like Canadian truckers on steroids, these supposed hayseeds gave the world a lesson in determination, ingenuity, courage, and organizational skill beyond the wildest dreams of the appalling bureaucratic yahoos who lord over them and seek to drive them to extinction. Rumors of French President Emmanuel Macron avoiding Paris hinted on possible lasting effects for the better.

In their protests, the farmers also displayed several of the highest of human character traits, including an admirable level of restraint against violence, and even a wicked sense of humor. It was inspiring and hilarious all at once. Watching them block roadways to major cities for weeks, simply "off-roading" in their tractors when confronted by the so-called authorities, was awesome.

When the farmers sprayed tons and tons of manure on various government buildings (talk about gilding the lily!) two questions entered my mind.

My first question, borne in part out of sympathy for the poor workers who would have to clean up the mess, was:

When scrubbing layer upon layer of bullshit off the halls of government, when does one stop?

My second question, more process-oriented, I suppose, was:

What permanent change will come from all this?

The subsequent actions of the French National Assembly on Valentine's Day answered my questions.

To my first question, the answer is: *never stop scrubbing.*

To my second question, the answer is: *nothing.*

On February 14, the French National Assembly passed **article 223-1-2 du code pénal**. Contained therein, in Article 4 of that law, Robert Kogon writes:

Article 4 introduces a new crime into the French penal code: incitation to abandon or refrain from medical treatment or to adopt a would-be treatment, if, "in the current state of medical knowledge", doing so "clearly" may cause harm to the person or persons in question. This crime is made punishable by one year in prison and a fine of €30,000 (£26,000) or, if the "incitation" has effect, i.e. the medical advice is followed, three years in prison and a fine of €45,000 (£39,000).

Kogon notes that this must pass the French Senate to become law. Still, it is an extremely ominous piece of legislation that clearly criminalizes medical dissent.

In effect, this is an extreme gag order on physicians, other health care personnel, and indeed anyone who dares speak out against official medical orthodoxy. In terrifyingly broad wording, it criminalizes – with hard time and crippling fines – advising against the received medical wisdom, even if the advice is not followed.

It does not take a doctor, lawyer, or medical ethicist to imagine the effect this will have on medical practice. Put simply, *this law will destroy the doctor-patient relationship.*

Throughout Covid, it became apparent how compliant and complicit the medical profession is to pressure from above. Doctors have been revealed to be a highly conformist bunch. This is understandable (although not excusable) given the nature of their training, professional conditioning, and employment structures.

With this law on the books, the few non-conformists must wonder, every time they advise a patient or place an order contrary to any "official" vaccine schedule, society practice guideline, or hospital protocol, if they will be reported to the authorities, and subject to criminal conviction, prison time, and huge financial penalties.

In the wake of Covid, this legislation demonstrates a blatant, damn-the-torpedoes attitude toward medical freedom. The Macron government has apparently learned nothing from Covid, save for adapting its excesses as templates for further governmental power grabs.

In the wake of the farmer's protests, it looks something like a test balloon. Massive, extremely well-organized protests by the farmers did reportedly win them some concessions. A rational person would think such civil unrest would also have chastened the French government from immediately attempting another outrageous assault on civil rights. Perhaps the government is just too stupid to see the connection. After all, what do farmers have to do with doctors?

Fortunately, brave activists like Annie Arnaud (@arnaud_annie26) in France and Kat Lindley (@klveritas) in the US, among others, have brought the issue to the forefront worldwide.

Will French physicians fight Article 4? Will ordinary Frenchmen fight it? For medical freedom, and for the doctor-patient relationship, this is a watershed case. The impact on French society will be profound and pernicious, perhaps even beyond the intentions of the wicked fools who are pushing it.

If Article 4 becomes law, the French government will have openly declared itself as totalitarian. The effects will ripple across Europe. For centuries, even long before the European Union, the fate of Europe has often been like a chain of dominoes, with France or Germany usually the first one tipped over. Can France – and Europe – be saved? Or was Burke indeed prophetic when he wrote, way back in the 1790s, that

...the age of chivalry is gone. That of sophisters, economists, and calculators has succeeded; and the glory of Europe is extinguished forever.

To those cleaning up after the farmers' protests, I offer one simple bit of advice:

Never stop scrubbing, *mes amis*. Never stop scrubbing.

Why Are Health Care Students Still Forced to Get Covid-19 Boosters?

Originally published on February 28, 2024 in Minding the Campus, and as "Health Care Students Still Suffer Force" *on February 29, 2024 in Brownstone Journal.*

A tremendous injustice is taking place in health care education, and most people are entirely unaware of it.

Today, almost four years since the Covid pandemic began, nearly all US medical students, nursing students, and students training in other healthcare fields are still being forced to choose between accepting continual booster doses of the Covid mRNA vaccines or being kicked out of their training programs.

This remains so, even though many institutions that enforce these mandates on health care students do not do so for faculty, staff, and patients.

This remains so, despite the fact that among the nearly 4,000 colleges and universities in the US, only 67 still require Covid vaccination for their undergraduate students—and even some of those holdouts do not require boosters. However, many of these same institutions that have rightly dropped mandates for their general student population still mandate Covid vaccination and boosters for health care students.

This injustice needs to end.

First, it is outright discrimination. It is unconstitutional, unlawful, and wrong. No mandate, especially one requiring submission to an invasive medical treatment, should be made on the basis of an individual's age, level of education, or rank in an organization. Health care students must enjoy equal protection under the law, equivalent to all others working in medical schools and hospitals.

Second, it does not stop disease spread. By now it has been firmly established—without further argument from vaccine manufacturers or the CDC—that Covid mRNA boosters do not create sterilizing immunity for individuals, and do not produce a herd immunity effect for the population. In fact, the CDC's own website makes no mention whatsoever of prevention of contraction or transmission of Covid in its description of "Benefits of Getting a Covid-19 Vaccine."

Put simply, if I compel you to take a vaccine that neither stops you from contracting the disease nor stops you from transmitting the disease, this will not protect me from the disease. Forcing medical and nursing students to take repeated Covid boosters does not protect patients.

It does, however, endanger students.

The dangers of repeated Covid boosters, especially in adolescents and young adults, are being increasingly acknowledged. The risks of vaccine-induced myocarditis and other severe and even deadly side effects are real and significant. Mandating repeated boosters at this late date, in an age group with a Covid case-fatality rate of less than 1 in 30,000 is wrong. The risk-to-benefit ratio is not even close to being favorable.

So why are Covid vaccines and boosters still being mandated for health care students?

Ask that question, and you are met with the same circular-finger-pointing excuses that shut down schools during the pandemic. No one claims responsibility, but everyone permits and promotes the injustice.

Even more troubling, health care students are commonly subjected to a cruel and dishonest game of bait-and-switch. According to the student advocacy group No College Mandates, "a healthcare student can secure a [vaccine] exemption for enrollment to study for a healthcare degree at the University of Pennsylvania or the University of Pittsburgh, but that same student cannot be placed in clinical rotations…unless they show proof of updated Covid vaccinations."

When confronted, the universities often blame the clinical training sites with which they are affiliated. However, most schools do little or nothing to accommodate the students to whom they *themselves* granted exemptions, such as finding clinical sites that do not mandate boosters.

Again, according to No College Mandates, one California State University department chair even declared "until 100% of [our] clinical sites drop the Covid vaccine requirement, our department will still require it."

Clinical sites, in turn, commonly cite local or state statutes—often vaguely or inaccurately—to justify their policies. John Coyle, attorney for a class action suit against Rowan College in New Jersey, characterizes schools blaming their clinical partners as a "shell game."

There is likely an underlying and utterly non-medical reason that these mandates persist. A covert screening process, often used in large corporations' Human Resources departments, appears to be taking place—this is an effort to weed out any and all individuals who do not passively comply with all regulations, however invasive or unnecessary they may be.

Such an approach poses grave dangers to the profession of medicine and to patient care. The history of medical progress, especially when it comes to good patient care, is filled with examples of reformers who fought harmful medical orthodoxy—and who were initially vilified.

"Weeding out" independent minds who question convention in favor of submissive, incurious drones will have disastrous effects on patient care.

If a mandate does not apply to everyone, it should not apply to anyone. This is fundamental to equal protection under the law in the US.

Practically speaking, these institutions should immediately drop these unjust, unconstitutional, and unhealthy mandates for their own good as well as that of their students. The Covid pandemic is over. No Covid emergency exists. Institutions that persist will be held accountable over time, and the legal jeopardy they place themselves in by continuing these mandates is potentially immense.

Health care students must take note of the unnecessary risks their universities are unjustly imposing upon them, gather together, speak out, and demand these mandates be dropped immediately and permanently.

Elected officials must take action to eliminate this and other residual injustices of the Covid pandemic, and to pass legislation to prevent such illegal overreach in the future.

Individual citizens must express their concerns to their elected officials and to the institutions where they receive health care.

The Covid catastrophe did immense damage to medical care, much of it the result of gross mismanagement at the highest levels of the industry. Those just entering it must be treated with renewed respect and consideration if they are to correct the mistakes of their predecessors.

Ending this injustice is an excellent place to start.

The Crucifixion of Kulvinder Kaur

Originally pushed on March 21, 2024 in *Brownstone Journal* and on March 26, 2024 *in The Epoch Times.*

I saw the best minds of my generation destroyed by madness, starving hysterical naked,
dragging themselves through the negro streets at dawn looking for an angry fix...
...expelled from the academies for crazy & publishing obscene odes on the windows of the skull...

ALLEN GINSBERG, *HOWL*

I think I finally understand the full meaning of the famous desperate words of the Beat poet quoted above. Allen Ginsberg's *Howl* laments the decimation of his generation, in part due to drug abuse and mental illness. (That much I got, even as a college kid.) But I have always suspected something more profound lay buried inside *Howl*, something beyond my comprehension.

Decades after its publication, it was revealed that much of the mayhem described in *Howl* was inflicted upon Ginsberg's generation by – you guessed it – its own government. Numerous 1960's countercultural figures such as Ken Kesey and Robert Hunter were survivors of the CIA's illegal and evil MK-ULTRA mind-control program, which of course was also the genesis of the 60s LSD culture as a whole. Other purported MK-ULTRA casualties, whose career paths turned very dark indeed, included the likes of Charles Manson, Whitey Bulger, and a teenaged Harvard undergraduate named Ted Kaczynski – later infamous as the Unabomber.

We cannot know to what extent Ginsberg may have been aware of the government's role in the generational destruction he described in *Howl* at the time he wrote it. But the deeper theme, the one that eluded me for so long, goes even further, and is more intuitive than factual. Those "best minds" were destroyed – and cast out of the academy – not by their *own* madness, but by the madness of the society around them. That society was a violent and unmerciful one, run by amoral and unaccountable men. It was a society that refused to accept alternative viewpoints and demanded conformity and submission. When the best minds failed to comply, it annihilated them.

A variation on this theme is playing today.

As a physician, I see the best of my generation being destroyed as well. Their careers are being stolen from them. They too are being expelled from the academies. They too are being destroyed by madness, but not their own madness, rather the madness of the society in which they live, of the profession in which they practice, and of the wicked and unscrupulous people in control.

There is a great purge occurring in the medical establishment, a purge that runs along strict ideological and ethical lines. The issue of Covid-era "misinformation" is the main pretext for this purge, but there is no reason to believe it will stop there. And although the American medical system is the most Pharma-captured and Deep State-riddled medical system in the world, this purge is by no means limited to the United States.

The list of honest, courageous, and self-sacrificing physicians and scientists who have been fired, censored, delicensed, defamed, subjected to lawfare, or otherwise persecuted in the name of Covid conformity is far too long to list. Just a very few of the names include Peter McCullough, Meryl Nass, and Martin Kulldorff in the US, David Cartland and Ahmad Malik in the UK, and Kulvinder Kaur in Canada.

Dr. Kaur faces imminent financial ruin at the hands of the Canadian court system, which has imposed a punitive $300,000 'cost order' upon her, due by the end of March 2024. This is in addition to other legal expenses she has incurred since the beginning of the Covid lockdowns.

Dr. Kaur's cardinal sin was speaking out against the harsh lockdowns imposed upon citizens of Ontario, where she practices medicine, treating mostly immigrant families and other poor members of the population.

Stanford epidemiologist and Great Barrington Declaration co-author Jay Bhattacharya recently interviewed Dr. Kaur on his Illusion of Consensus podcast. I encourage readers to watch this interview. It is compelling for a number of reasons, not the least of which is this: Kulvinder Kaur comes across as the most modest, sincere, and likable person imaginable – quite literally the last person who would invoke the ire of any honest organization, and quite possibly the first person you would want as your personal physician.

I do not know Dr. Kaur personally, though I am pleased to say I know Jay Bhattacharya. And as anyone who knows Jay will affirm, he is a delightful person, truly a warm and generous soul.

Consider the fact that I recently attended a talk given by Jay. It frustrated me to no end. Why? Jay's topic: finding empathy for Anthony Fauci.

Don't forget, Jay was one of the "fringe epidemiologists" against whom Fauci and Francis Collins ordered a "swift and devastating takedown." Only a person of tremendous kindness and forbearance (much greater than I possess, that's for sure) can love his enemies like that.

Anyway, in their interview, Kulvinder Kaur makes even Jay look like a character out of *The Sopranos*. Her combination of earnestness and modesty shines through her descriptions of both her first class education and scientific training and her clinical practice, devoted as it is to poor immigrants, in large part because she herself is an immigrant. Prior to Covid, she says, she was a very conventional practitioner, observant of the standard vaccine schedules, and never in trouble with the authorities.

But when the lockdowns began, she felt compelled by conscience to speak out against the collateral harms these repressive measures caused her patients. She cited sources such as the Great Barrington Declaration. She took to Twitter. She refused to shut up. And so, the Canadian establishment set out to destroy her.

In her interview with Jay, Kulvinder Kaur appears to me to be very idealistic, sometimes almost to the point of naivete. At one point she states, "I never thought that this would be the cost of speaking truth to power." She seems genuinely surprised that even now, when her predictions from early in the pandemic have all been proven correct, she is still being persecuted.

The ideologues are hellbent on destroying the idealist, and the idealist cannot understand why.

But persecuted she is. The Canadian government, media, and medical establishment long ago determined that it would publicly crucify (and financially ruin) this highly intelligent, deeply moral, and utterly sincere young woman. They intend to make an example of her, just in case some other idealistic young physician is thinking of following in her footsteps.

A GiveSendGo fund for Dr. Kaur's financial crisis has been started, and I encourage readers to donate to it as soon as possible if they are able. She needs to raise $300,000 by the end of March. Hopefully the goal will be reached, and Dr. Kaur will be saved from financial ruin.

But the ruin of medicine in the so-called Western democracies will continue apace. The best of my generation of physicians will be destroyed by the madness of their corrupt, captured, and deteriorating profession. And then where will patients turn for care?

What parent, when seeking a physician for their child, would not choose an impeccably trained, self-sacrificing clinician who has devoted her career to the care of the poor, over some vindictive, venal, Machiavellian technocrat? Put another way, what Ontario parent would not choose Kulvinder Kaur over say, the execrable David Fisman?

If the current trend of ruining the careers of honest, courageous physicians and scientists is not stopped, such choices will become, for want of a better word, academic. The outstanding, outspoken, and independent physicians will be run out of the profession. The remaining rank-and-file, already more submissive than their persecuted betters, will quietly comply with orders from above, knowing what will happen to them if they don't. The newly minted doctors, freshly indoctrinated in today's Pharma-driven curricula, and pre-selected for compliance via mandatory vaccinations and other human resource department litmus tests, will goose-step through their practice directives and clinical protocols, no questions asked.

(Footnote: the publication and sale of Howl *resulted in the arrest of its publishers and culminated in a famous 1957 legal case involving – you guessed it – censorship and the First Amendment.)*

Crush the Flu d'État!

*Originally published on May 6, 2024
in* Brownstone Journal.

As a practicing physician, I have had numerous conversations over the years with people caught in dysfunctional relationships. While it is often best to refrain from "telling" such persons the "right" course of action, one notable exception is when the relationship is clearly abusive.

The cure to an abusive relationship is straightforward, if sometimes difficult: leave.

The World Health Organization's (WHO) relationship with member nations, including the United States, is a classic example of an abusive relationship.

According to the US Health and Human Services' Office on Women's Health, an abusive relationship is characterized by a partner who:

- Controls what you're doing.
- Checks your phone, email, or social networks without your permission.
- Decides what you wear or eat or how you spend money.
- Prevents or discourages you from going to work or school or seeing your family or friends.
- Humiliates you on purpose in front of others.
- Threatens to report you to the authorities for imagined crimes.

What is the abuser's motivation for such acts? According to the National Domestic Violence Hotline, "One feature shared by most abusive relationships is that the abusive partner tries to establish or gain power and control through many different methods at different moments."

Sound familiar?

If not, please take the time to do the following.

First, recall what the entire world was subjected to, starting around the Ides of March 2020:

- Gross civil rights offenses creating intense personal isolation and financial distress (lockdowns).
- Advanced psychological manipulation techniques creating fear, uncertainty, and dependency ("social distancing," forced masking, "distance learning," and endless "fear porn").
- Egregious, population-wide violations of medical ethics amounting to industrial-scale physical assault (coercion and mandates to make millions accept repeated doses of the experimental Covid vaccines).

Second, read the WHO's proposals (see footnotes). Even in their "revised" form, the WHO seeks carte blanche to repeat the whole process, entirely at their own discretion.

Next, read Dr. David Bell and Dr. Thi Thuy Van Dinh's critique (also in footnotes) of these highly deceptive documents, which represent a blueprint for, you guessed it, the continued *abuse* of free people on a global scale. Bell and Dinh's highly persuasive argument: the changes made to the WHO's pandemic proposals, as a result of extensive pressure, are "merely cosmetic." In other words, the WHO is concealing its real intent.

I will not dilute their detailed criticism here. I will state that they carefully outline, among other problems, the highly deceptive language, the immense potential for corruption, and the fundamental epidemiological fallacies contained in the WHO's proposals.

Furthermore, please note that the WHO is absolutely not "who" they say they are. By the WHO's own accounting, the single largest contributor to its coffers is Bill Gates. The Gates Foundation and Gates-controlled GAVI provide over 20% of the WHO's funding.

Finally, recognize that after making its mildly watered-down revisions, the WHO is breaking its own rule that requires a 4-month minimum period before member countries vote on new proposals. Despite the revisions, the WHO is insisting on the original May 2024 deadline. The WHO is evidently in a great hurry to rush its latest lipstick-coated pig to market.

Let's put it another way. Imagine that you and I are partners of some kind: domestic partners, business partners, whatever. I attempt to impose a complex legal agreement upon you. This agreement empowers me to control your freedom, money, and even your bodily autonomy, in the event that a hypothetical emergency occurs (which, by the way, I can declare at any time). You read the document and say, "That's crazy!" So I water it down a little, in highly deceptive ways, throw the new version back at you, and give you no additional time to review it.

What would you do in that scenario?

If you possess a lick of sense, you would tear the agreement up like Nancy Pelosi with a State of the Union address. You would throw the shreds of paper in my face. You would walk away and have nothing more to do with me.

In the wake of the manufactured Covid-19 catastrophe, pandemic preparedness has become the preferred fear-mongering and power-grab tactic of the global elites and the military-medical-industrial complex. The WHO is a central figure in that tyrannical cabal.

The inimitable Ivor Cummins has dubbed the WHO's pandemic power grab proposals as the "flu d'état." This brilliant pun/neologism describes the WHO's intent perfectly and succinctly: it means to use the threat of disease to illegitimately seize governmental power.

No way. We must mercilessly crush the WHO's attempted flu d'état.

What then, you may ask, do we do about potential pandemics, once we leave the WHO?

First, we must recognize that any risk of future pandemics comes overwhelmingly from human manipulation of pathogens, rather than natural pathogens, as sure as SARS-CoV-2 came from the lab.

Second, we must shut down every possible bioweapons laboratory, be it located at Fort Detrick, the University of North Carolina at Chapel Hill, Wuhan, or the Ukraine.

Third, we must put the Faucis, Daszaks, Barics, and Bat Ladies of the world in the dock, to be tried for crimes against humanity.

Fourth, we must reconstruct public health as a bottom-up network of communicating local entities, rather than a top-down tyrannical enterprise.

But those steps are, to varying degrees, difficult to achieve. Leaving the WHO is simple.

Many people, even politicians, are finally waking up to the WHO's abuse, and some are even doing something about it, largely thanks to the work of committed advocates like Drs. Bell and Dinh, Dr. Kat Lindley, the irrepressible Dr. Meryl Nass, and others.

In the United States, multiple state and local governments have declared that WHO policies will not stand in their jurisdictions. On May 1, 2024, a group of 49 Senators (all Republican) sent a letter to President Biden telling him to withdraw support for the WHO treaty and amendments.

They further warned that even if the US does proceed, such an agreement would constitute a treaty and thus would be subject to Senate review and would require a 2/3rds vote of the Senate to be approved.

All good. But once again, the definitive step goes beyond these measures.

It is not sufficient to renegotiate this treaty, agreement, or whatever you want to call it. It is not enough even to scuttle it completely and then recalibrate our relationship with the WHO. We must not waste time trying to reform this corrupt and illegitimate organization.

We must get out.

Part of the beauty of leaving the WHO is this: not only is it simple, it's easy. The WHO (like its manipulative, dysfunctional parent, the UN) is a paper tiger. The WHO has zero authority above and beyond that which we grant it. Unlike an unfortunate woman trapped in a violent household, the WHO cannot beat us up, steal our money, or kidnap our children. Not yet.

There is a cure for the abusive relationship we have with the WHO. It is the standard corrective for abusive relationships. The solution is not to negotiate, to reconsider, or to give the abuser one last chance. The solution is to leave.

We must leave the WHO.

Does "One Health" include assassinations?

Written with Joshua Stylman,
and originally published on May 29, 2024 in Substack.

On May 15, 2024, Slovakian Prime Minister Robert Fico was shot multiple times in an assassination attempt in the town of Handlova, Slovakia. The attack occurred in broad daylight after a government meeting.

In November of 2023, Mr. Fico had announced that Slovakia would not support the World Health Organization's (WHO)'s Pandemic Treaty nor its updates to the existing International Health Regulations, stating:

> *I also very clearly declare, that SMER Slovakian Social Democracy (the ruling political party) will NOT support the strengthening of World Health Organization's powers at the expense of sovereign states in managing the fight against pandemics.*

I will say that such nonsense could only have been invented by greedy pharmaceutical companies which began to perceive the resistance of certain governments against mandatory vaccination.

The attempt on Mr. Fico's life comes at a critical time for the WHO and its proposed Treaty.

Multiple nations have expressed disagreement with its contents, to the point where the WHO significantly revised the Treaty's wording, even renaming it an "Agreement." However, leading critics have called the revisions "merely cosmetic," support is uncertain, and the WHO has broken its own procedural rules in refusing to allow a 4-month period to review the new documents.

Within 24 hours of the assassination attempt on Mr. Fico, major Western news outlets including *The Guardian* and the *New York Times* were promoting a statement attributed to the country's interior minister that the shooter was a "lone wolf."

Needless to say, declaring a political assassin to be a "lone wolf" a mere 24 hours after the act is presumptuous in the extreme and strongly suggestive of a cover-up. No honest, truth-seeking news agency would lead with such a claim, as *The Guardian* did, or present that as the sole hypothesis, as did the *Times*, especially so soon after an assassination attempt.

Just a day later, *Bloomberg News* reported on May 16, 2024 that under the EU's new Digital Services Act, "The European Commission said it is "actively monitoring" the spread of fake news about Wednesday's shooting [sic] Slovak Prime Minister Robert Fico and warned it can slap Big Tech platforms with fines for failing to tackle disinformation."

Declaring oneself to be the arbiter of truth immediately in

the wake of such an act of sudden political violence, as the European Commission has done, is both arrogant and, once again, highly suspicious.

Unsurprisingly, just a few days later, pesky facts intruded that exposed the "nothing to see here" false assurances by the legacy media and the threats of censorship by international authorities as rank dishonesty and likely attempts at a cover-up. By May 19, *Bloomberg News* reported that

> *A potential broader assassination plot is supported by the fact **that the assailant's social media communications were erased by another person** about two hours after the shooting, Interior Minister Matus Sutaj-Estok told reporters on Sunday.*

Notably absent from the media's highly suspicious coverage of the attack on Fico has been any mention of the elephant in the room: Robert Fico's publicly declared stance on the rapidly upcoming WHO vote.

Had this assassination attempt on Prime Minister Fico been a unique occurrence, one might dismiss its relationship to Mr. Fico's statements about the WHO as coincidence. However, an alarming series of similarly suspicious assassinations and sudden deaths of world leaders has accumulated since the onset of Covid. All of these cases involve leaders who openly defied the WHO. It is no longer plausible to dismiss these as "coincidences."

The simple truth is this: since Covid, if a world leader openly defies the WHO, either outright assassination or death by highly questionable circumstances is likely to follow.

President John Magufuli of Tanzania

John Magufuli was the president of Tanzania from 2015 until his death in 2021. Magufuli, who held a doctorate in chemistry, was a highly skeptical and vocal critic of the WHO's directives regarding Covid-19. He rejected the clinical validity of Covid tests, even having a goat, sheep, and a paw-paw fruit swabbed for Covid and their samples sent to a lab, where they tested positive. He mocked the efficacy of masks, and refused vaccines for Tanzanian citizens.

In February 2021, WHO Director General Tedros Adhanom Ghebreyesus expressed concern over Tanzania's handling of the virus. The WHO repeatedly pressed Magufuli to "follow science," share data, implement their measures, and prepare for vaccines. WHO's Africa director reportedly called Tanzania's stance "very concerning" and a regional outlier.

A February 8, 2021 article in The *Guardian* stated "Magufuli's cavalier disregard of COVID's impact in the great lakes region is fueling conspiracies and endangering lives." (Under its title, that same *Guardian* article notes that it was sponsored by the Bill & Melinda Gates Foundation. Around that time, Gates contributed a sum of nearly $13 million to The *Guardian*. *equating to the sum of $116 per reader of the print version of the newspaper.*)

Magufuli's administration had also kicked Gates and Monsanto food initiatives out of Tanzania just a few years prior. Of course, the relationship between the WHO and Bill Gates is extremely close. By the WHO's own accounting, the single largest contributor to its coffers is – you guessed it – Bill Gates. The Gates Foundation and the Gates-controlled GAVI combined provide over 20% of the WHO's funding.

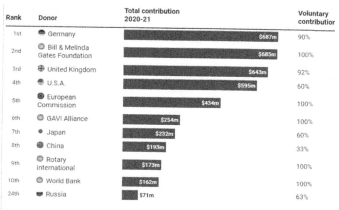

Figures include mandatory ("assessed") contributions, voluntary contributions and projected funding. Total committed funding over the two-year period could change if donors commit additional funding, for instance.

In March 2021, Tanzanian President John Magufuli died unexpectedly, with the official cause of death shrouded in uncertainty. While some speculated that Covid-19 may have been responsible, the circumstances surrounding his demise, including the timing and absence of verifiable firsthand accounts, cast doubt on this assertion. Furthermore, multiple sources, including The *East African* and Factcheck.org, attributed his death to "chronic atrial fibrillation."

However, chronic atrial fibrillation is a common and very treatable heart condition. While it can lead to other serious health complications, it is rarely an imminent cause of death in itself.

Immediately after Magufuli's death in March 2021, the new Tanzanian government, under new President Samia Suluhu Hassan, changed course and immediately began pursuing Covid-19 vaccines through the WHO's COVAX initiative and bilateral deals.

President Jovenel Moïse of Haiti

An eerily similar storyline transpired in Haiti. In May 2021, the Haitian government under President Jovenel Moïse refused approximately 750,000 doses of the AstraZeneca Covid-19 vaccine that were scheduled to arrive in Haiti through the WHO's COVAX initiative.

According to an unnamed government source cited by the Dominican news outlet EFE, Haiti had requested vaccines from other manufacturers instead. The specific reasons for refusing the AstraZeneca vaccines were not clearly stated, but it came amid global concerns about "rare" blood clots linked to the vaccine. This refusal of the COVAX vaccines contributed to delays in Haiti's vaccine rollout. Furthermore, President Moïse was quoted as stating that even when vaccines became available in Haiti, he would not mandate them.

Moïse was assassinated on July 7, 2021 when attackers stormed his private residence in the Haitian capital of Port-au-Prince.

On July 14, 2021, just one week after the assassination of President Moïse, the first vaccine shipment containing 500,000 doses of the Moderna vaccine arrived in Haiti through the WHO's COVAX program.

(Note: Last week, AstraZeneca finally pulled its Covid vaccine from the world market due to blood clot-related fatalities.)

President Pierre Nkurunziza of Burundi

Another concerning case is that of Burundi's ex-President Pierre Nkurunziza. Burundi was one of the few countries that did not implement strict lockdown measures or promote social distancing during the early stages of the pandemic. Like the other world leaders highlighted, President Nkurunziza also opposed draconian measures to combat the pandemic. Like the others, Nkurunziza died unexpectedly.

In May 2020, Nkurunziza's government expelled World Health Organization (WHO) representatives from Burundi, approximately a week before national elections were held.

International public health officials (and the opposition party) publicly criticized the decision to hold elections, even though there had been only several dozen reported cases of Covid in Burundi by that time.

The circumstances surrounding Pierre Nkurunziza's sudden death on June 8, 2020 have raised suspicions, given the timing and context of the pandemic. The official cause of death was reported as a cardiac arrest. It occurred just three weeks after expulsion of WHO officials and the controversial 2020 presidential election, which he did not contest after serving three terms.

The expulsion of WHO officials came after the WHO raised concerns about the government's handling of the Covid-19 pandemic, including the lack of social distancing measures and the continuation of large gatherings. The WHO had also accused the Burundi government of lacking transparency regarding the virus's spread within Burundi.

After Nkurunziza's death, he was replaced by President Evariste Ndayishimiye, who took a very different approach to the pandemic. On October 6, 2020, the International Monetary Fund (IMF) announced that it had reached an agreement in principle with the Burundian government on a $78 million aid package to help the country cope with the economic impact of the pandemic.

Just a day after this announcement, the Burundian government seemed to have a change of heart regarding its stance on how to deal with the pandemic. New President Ndayishimiye declared that Covid-19 was the country's "worst enemy" and called for increased vigilance and testing, as well as the acceptance of vaccines.

Prime Minister Shinzo Abe of Japan

On April 17, 2020, at the height of the first wave of Covid, Japanese Prime Minister Shinzo Abe told the media in a press conference that the WHO "has issues" and that Japan would review its financial contributions after the pandemic is over.

Abe did not impose strict lockdowns during Covid. A close ally of President Donald Trump, Abe also ended Japan's Covid Declaration of Emergency on May 25, 2020, much earlier than many other countries. (By comparison, the United States did not end its Covid State of Emergency until *three years later*, on May 11, 2023!) At the time Abe ended the Emergency, Japan had suffered only 850 Covid deaths among a population of 120 million people.

Nevertheless, Abe received great criticism from the mainstream press for doing so.

Abe resigned from office on September 16, 2020, citing a recurrence of a debilitating chronic colitis condition.

Shinzo Abe was assassinated in Nara, Japan on July 8, 2022, by a man using a homemade firearm. The assassin's stated motive was Abe's association with the Unification Church. Apparently the "lone wolf" struck again.

Fact or Fiction? Unmasking the Arbiters of Truth

Various "fact-checking" organizations have aggressively contested any speculation regarding the deaths of these leaders, attacking with extra vigor any theories implicating the WHO, Big Pharma, or worldwide vaccination. The approach of these self-appointed arbiters of truth appears to be either to sloppily proclaim to have debunked all such hypotheses *en masse*, or to merely dismiss them outright as "baseless conspiracy theories."

However, the impartiality of these "fact-checkers" is itself highly questionable, as they have financial incentives and industry ties that are concerning indeed.

For instance, on June 30, 2021, Reuters definitively stated 'These four leaders were not killed for opposing COVID-19 vaccination,' referring to the cases of John Magufuli, Pierre Nkurunziza, and two other African leaders who died around the same time (Ivory Coast prime minister Hamed Bakayoko and eSwatini prime minister Ambrose Dlamini).

From its deceptively narrow title on down, the Reuters article reads like an object lesson in sloppy, tendentious journalism. It's unclear how much original investigative reporting, if any, the "fact-checkers" performed in the process of drawing their blanket conclusion. The article confusingly conflates the cases of two leaders who were highly defiant of WHO (Magufuli and Nkurunziza) with two others who were not.

Finally, the article brandishes definitive statements like "There is no evidence that he [Magufuli] was murdered," and "There is no evidence that Nkurunziza was killed" without supporting evidence or citation, and they fail to mention the rapid shifts in national policy that occurred in both countries *toward WHO's directives* following the leaders' respective deaths.

Perhaps it also would have been pertinent for the fact-checkers to mention the undisputed fact that James Smith, their longtime CEO through March 2020, sits on the Board of Directors at Pfizer.

One of Factcheck.org's intrepid truth-seekers, one Brea Jones, even went so far as to dismiss suspicions surrounding President Moïse's assassination, a mere *one week* after his murder, as "baseless conspiracy theory." This bold claim was made while within the text of the article, she acknowledges:

"While *the investigation is underway*, no evidence has been surfaced [sic] suggesting that Moïse was assassinated because of opposition to the COVID-19 vaccination."

Apparently, for Brea Jones and Factcheck.org, one week is enough to proclaim gospel truth about the circumstances surrounding the assassination of a head of state, *even while the investigation is still underway.* Where was Miss Jones when the Warren Commission needed her?

Investigation into the funding of the prolific Factcheck.org reveals that Factcheck receives funding from both the Robert Wood Johnson Foundation (the charitable Foundation of pharmaceutical giant and Covid vaccine manufacturer Johnson & Johnson) and the Annenberg Foundation (who, in turn, receives funding from The Bill & Melinda Gates Foundation).

PolitiFact, another prominent fact-checking organization whose stated mission is to "publish the truth so you can be an informed participant in democracy," also addressed this controversial claim. However, like some other fact-checkers reporting on this issue, PolitiFact did not disclose in the article that its parent organization, the Poynter Institute, has also received significant funding from the Bill & Melinda Gates Foundation.

Of course, Bill Gates has been one of the world's most vocal advocates for vaccines, which by his own admission hasn't been an entirely altruistic endeavor. In a 2019 interview, he described his investments in global vaccination efforts as yielding a 20-to-1 return in economic benefits, calling them his "best investment" ever.

Gates' financial support of media outlets like PolitiFact (and numerous other major news outlets) creates at minimum a perceived conflict of interest when those outlets report on vaccine-related controversies.

The highly questionable associations and funding sources of so many of the self-appointed "fact-checkers," coupled with their habit of rushing headlong to prematurely dismiss "conspiracy theories," would be comical were they not so pernicious. Assassinations are intrinsically conspiratorial in most instances, and it should be the sworn duty of an *honest* press to seek out the underlying causes of such destabilizing events, not to simply smother inquiry, especially in the favor of those who cut their paychecks.

Fatal Defiance: The Perils of Challenging the WHO

Coincidences do occur, but not in clusters of five. How plausible is it that every one of these "coincidences," involving the sudden deaths of heads of state – three of which were undisputedly assassinations or assassination attempts – were unrelated to the victim's well-documented public defiance of the WHO?

How plausible is it that after at least 3 of these "coincidences" occurred, the leader's death prompted an immediate *volte-face* in his country's health policy, turning it strongly in favor of the WHO's wishes?

How plausible is it that, if these were indeed "coincidences," the usual flotilla of compromised, tendentious "fact-checkers" had to be repeatedly deployed to debunk any questioning of these inherently suspicious and deeply troubling events?

With the vote on the WHO Pandemic Treaty rapidly approaching (despite eleventh-hour changes in language and the breaking of its own procedural rules on time for discussion and review of the documents), many people worldwide are speaking out against the treaty. This includes thoughtful, sincere, and valid analysis from smaller and developing countries.

For example, The Pan-Africa Epidemic and Pandemic Working Group, a collection of senior academics from seven African countries, released a statement to the African Union. It calls for a halt to the upcoming Pandemic Treaty and IHR amendment votes on several grounds, including conflicts of interests, the WHO's poor track record during Covid, and the procedural irregularities and nondemocratic nature of the entire process.

How likely is it that leaders in African nations, smaller countries, or anywhere else will listen to their own people and their own best advisors if the specter of assassination hangs overhead?

Nations around the world should reject the WHO Pandemic Treaty and changes to the IHR. In fact, the WHO should be disbanded entirely. The WHO has amply and repeatedly demonstrated itself to be a shady, dishonest, and power-hungry entity. The well-founded suspicions of political assassinations of leaders opposing this organization are but one reason for this distrust, but one that must be emphasized at this critical time.

Questioning Modern Injection Norms

>Originally published June 14, 2024 in *Brownstone Journal* and in *RealClear Health*.

A recent medical study found an association between tattoos and malignant lymphoma, with a 21% increased risk of this type of cancer in tattooed persons. Published in the *Lancet* (oh, the irony!), the paper notes that tattoo ink contains known carcinogens. Nevertheless, the popularity of getting inked has skyrocketed in the past few decades.

Within living memory, the idea of having things injected into one's body was generally viewed with aversion. The horror of intravenous drug addiction and the specter of AIDS both played a role in this. Still, there is a natural terror of having one's skin penetrated that is – or at least was – inherent to the human psyche: consider the enduring popularity of vampire mythology as a staple of the horror genre.

Children in particular have always had a hatred of needles, and for good reason: first, it's an obvious invasion of their physical person, and second, it hurts. Holding down a struggling child to inject them with a vaccine (often while insisting to them it's for their own good) is a perennial litmus test for medical students as they decide upon their specialty of choice. After all, if you're not willing to overpower young children and force needles through their skin, you'll have a hard time making a living as a pediatrician.

In my estimation, human distaste for the hypodermic route of administration is both perfectly natural and adaptive to survival. The skin is the body's largest and most important barrier to infection and injury, and any breach of it is potentially dangerous.

In nature, who tries to penetrate our skin? Parasites, poisoners, and predators, that's who. Mosquitoes and other biting insects. Blood-sucking leeches. Stinging insects like hornets and wasps. Venomous animals, especially snakes. Large predators that will eat you if they can, from big cats to crocodiles to sharks.

And of course, other humans with their weapons.

In nature, the consequences of having one's skin pierced are serious and potentially deadly.

Obviously, large-scale hemorrhage can result in death. However, dangerous infections of many kinds can also result from even a small breach in the body's integument.

For example, malaria, an infectious disease caused by a single-celled animal (protozoan), and still a leading cause of death in the developing world, is contracted via mosquitoes. Lyme disease, caused by the probably laboratory-altered bacteria *Borrelia Burgdorferi* and ubiquitous in the United States, is transmitted by tick bites. More mundane perhaps, but just as dangerous, virtually any open wound, if neglected, can become infected by numerous bacteria – or even fungi – and result in sepsis and death.

So why are we so eager to have our skin penetrated these days? Tattoos, body piercings, injection pharmaceuticals, and of course vaccines are all much more prevalent today than even a few decades ago.

Tattoos not only are much more common today, they are also much more extensive, often covering entire limbs, or even entire people. I have yet to diagnose a case of tattoo-induced lymphoma, but I have seen several nasty cases of tattoo-induced cellulitis, and in the old days, Hepatitis C infections with no other known risk factor.

Body piercings have followed the same pattern as tattoos: more of them and more extreme examples. Ears with 10 earrings each. Nose piercings, both in the nares and the septum. Eyebrows, lips, tongue (it enhances certain types of sexual stimulation, or so I have been told), nipples, navel, and of course, genitalia. And I'm sure I am forgetting something.

Today, many commonly used drugs are injection-based. Numerous immunomodulators for autoimmune diseases are given by injection, such as Humira, Enbrel, and Skyrizi, among others. Some have black-box warnings for life-threatening side effects. They sell like hotcakes anyway.

Injection hormonal medications such as anabolic steroids and Human Growth Hormone (HGH) are frequently used – and misused – to promote muscle growth, enhance athletic performance, and prolong youthfulness. Conversely, testosterone suppressors such as Lupron are injected into prostate cancer patients *and* men wishing to transition into women.

Insulin has been around for about 100 years, and for most of that time, it was the only injection medicine for diabetes. Nowadays, following the explosion in prevalence of type-2 diabetes, a number of new injectable diabetes medications have reached the market. They have proven extremely popular (and profitable) and are now being used for non-diabetic diagnoses as well, most notably for weight loss.

The diabetes drug semaglutide has become so popular as a weight loss treatment that

- It goes by three trade names (Ozempic and Wegovy are the injectable versions. An oral preparation is known as Rybelsus.)
- It has transformed its manufacturer, Novo Nordisk, into the most valuable company in Europe, with a market capitalization bigger than the entire economy of its native country of Denmark.
- Its availability hampered by intense demand, a black market has developed around the so-called "skinny jab."

To summarize the current state of injectable medications: if you're a man, and you want to be more of a man, there's a shot for that. If you're a man, and you want to be a woman, there's a shot for that. If you're a fat man and you want to be a skinny man, there's a shot for that, too.

Last but not least, there are the vaccines.

Since the National Childhood Vaccine Injury Act of 1986 (NCVIA) was signed by President Reagan, forever protecting vaccine manufacturers from liability, there has been a dramatic increase in the number of vaccines brought to market. This is reflected in the number of vaccines added to the CDC vaccine schedules, with the number of vaccines on the CDC Child and Adolescent schedule rising from a mere 7 in 1986 (how lucky we were!) to a whopping 21 in 2023.

The Covid mRNA injections have upped the ante for repeated jabs – to an extreme degree. Some patients who actively sought every recommended booster dose over the last three years have received 6 or 7 total Covid shots by now.

Big Pharma clearly views the mRNA platform as a plug-and-play model for numerous new medications. Furthermore, while actually gene therapies, the mRNA products are actively billed as "vaccines" to keep them under the NCVIA anti-liability umbrella.

On its own website, Moderna describes a pipeline of mRNA vaccines currently in development for Influenza, Respiratory Syncytial Virus (RSV), Cytomegalovirus (CMV), Epstein-Barr Virus (EBV), Human Immunodeficiency Virus (HIV), Norovirus, Lyme disease, Zika virus, Nipah virus, Monkeypox, and others.

With the current H5N1 alarmism currently in play, promoted by Covid figures such as Deborah "Scarf Lady" Birx, the game plan is clear.

Covid was not an aberration. Covid was a dress rehearsal.

By the way, many vaccines contain aluminum. Aluminum is an established neurotoxin. But don't worry, Mommy. Kids are resilient, remember?

Many vaccines contain thimerosal. Thimerosal is a mercury compound. Mercury is an established neurotoxin – the cause of the Mad Hatter's madness, as mercury was used in the making of felt. Long before Lyme, Connecticut became famous for its eponymous disease, hat-making center Danbury, Connecticut was known for the "Danbury Shakes."

But don't fret, Mother. Vaccines are safe and effective by definition, remember?

Patients were told that the Covid mRNA injections did not contain potentially carcinogenic SV40 DNA. Of course, now we know they are contaminated, and as cancer diagnoses increase, especially in the young, patients are told, just as they were regarding myocarditis, to believe the 'experts' rather than their own lying eyes.

But the final frontier of jab-happiness has arrived when the pregnant woman is invited to the party.

Historically, pregnant women were universally and correctly viewed across medicine as extremely vulnerable to iatrogenic (treatment-induced) injury. As a result, they received maximum protection from it – meaning they received the absolute minimum possible treatments and interventions.

To this old-fashioned – or perhaps just old – doctor, the fact that pregnant women are now recommended to receive both the Covid-19 mRNA jabs and the new RSV injection is proof positive that:

- The pre-Covid standard of *primum non nocere* ("First, do no harm") in medical ethics is dead and buried. Let the buyer beware.
- The priority of the medical industry must be assumed to be the promotion of an agenda, policy, and/or product, rather than the well-being of the individual patient, until proven otherwise.

Before I am accused of calling for the outlawing of all hypodermic needles and the banning of every parenteral medicine, I will clarify both what I am saying and what I am not.

Certainly, there are legitimate uses for injection medicines. An obvious example: countless type-1 diabetics have been able to live full lives due to the presence of insulin in the medical pharmacopeia. Were injection insulin not available, many millions would have died over the past century. Similarly, intravenous medicines have saved many millions as well, especially critically ill and hospitalized patients.

There is unquestionably a role for injection medications. But there are risks and harms, both known and unknown, to their use. The current mindset, which appears to be 'If there's a medical issue, there's a shot for that,' is deeply problematic.

A shot in the arm is to some extent a shot in the dark. In general, the 3 most common types of non-intravenous injections are intradermal, subcutaneous, and intramuscular. With proper technique, a skilled doctor or nurse can do whichever one is called for, with a high degree of precision.

However, potentially harmful mishaps do occur, such as accidental intravascular injection (directly into a blood vessel). Having inexperienced and/or minimally trained persons such as pharmacists, pharmacy assistants, medical assistants, and even totally non-medical persons perform injections, as happened widely during Covid, increases the risk of complications.

Perhaps the most dangerous aspect of this injection-based mindset of health care is the flawed view of reality it creates. The obesity epidemic is caused by excessive calorie intake, grossly unhealthy diet, and lack of physical activity. It is not the result of a population-wide Ozempic deficiency.

We have an immune system for a reason. The human immune system has served our species well throughout our existence on earth. It is competent, capable, and breathtakingly complex – far beyond the understanding of Anthony Fauci and Stephane Bancel, I should add. It does not help the immune system, or us, to hyperstimulate it dozens and dozens of times during childhood with injection after injection, only to suppress it in later life with even more injections when it has been driven haywire.

The human immune system does not need a crude, laboratory-made primer for every antigen it faces. I know there's no money in this approach, but nevertheless: leave it be. Let it do its job.

Likewise, we have skin for a reason. It is present to shield the interior of our bodies from harmful elements in the outside world. When we violate that shield, we subject ourselves to dangers that are obvious (such as bleeding) and to dangers that are invisible (infections, toxins, and assaults on the immune system). If you don't think the skin is a complex immunologic organ, just ask any piercing enthusiast with a nickel allergy, or better yet a few of the Covid jab recipients who developed Stevens-Johnson Syndrome.

The current, extremely blasé attitude of today toward this important aspect of bodily integrity, as promoted both by Big Pharma/Big Medicine and our culture as a whole, is a big mistake.

The natural routes of entry into the human body, be they for food, air, or reproduction, do not include penetrating the skin. This mode of introduction of foreign material is inherently unnatural, abnormal, and potentially dangerous. When truly necessary and properly performed it should be used, but when unnecessary it should be avoided.

When you recoil at the thought of a needle penetrating your skin and injecting you with something, this is a normal, sensible, and self-preservatory reaction. You may note that this aversion to needles is similar to how you would feel about a mosquito, a leech, a snake bite, or even a knife in your back. This is not a coincidence.

Parasites, poisoners, and predators come in many sizes, shapes, and species. Become as knowledgeable as possible about anything you allow to be done to you. Listen to your own God-given body. Trust your own instincts. Learn to say no.

Protect your bodily integrity. Protect yourself.

Pandemic Preparedness: Arsonists Run the Fire Department

Written with Brian Hooker, PhD.
Originally published on July 1, 2024 in *Brownstone Journal*
and on July 2, 2024 in *The Defender of Children's Health Defense*.

Imagine if you will, an exceptionally ambitious city Fire Department, located in a city with very few naturally occurring fires.

These ambitious firemen don't have nearly enough work, prestige, or pay for their liking. Uninterested in simply polishing their trucks, lifting weights, and cooking chili, these firemen want more. A lot more.

They construct a plan. They will start a research program, funded by taxpayers, whereby they will develop an arsenal of the biggest, scariest, most flammable products on earth. They will justify this program under the pretense that these destructive creations are absolutely necessary for the development of bigger and better fire *extinguishers*. Incidentally, they will also develop, market, and sell these fire extinguishers themselves.

These proprietary fire extinguishers will net the ambitious

firemen an incredible fortune – if they can just get every man, woman, and child in the city to buy one.

The Fire Department, working with the corporations that would manufacture their miracle extinguishers, actively publicizes the supposedly tremendous, ever-increasing risk of fires that they claim threaten the population. According to the ambitious firemen, risk factors for worsened fires are everywhere and are ever-increasing – global warming, population growth, take your pick – and the next "big one" is just around the corner.

Credulous, fearful citizens and heavily lobbied politicians fall for their story, pumping ever more tax dollars into the Fire Department's research and development program.

The Fire Department develops and grows its stockpile of manufactured fire super-hazards, until one day…

OOPS!

Somehow, one of the flammable products is released, and a raging conflagration ensues. No one knows exactly how it started – in fact, the chief firemen gather together and publicly deny that any of their products could be responsible.

But by terrifying the public and confusing the politicians, the firemen coerce the population to shelter in place and follow their strict instructions, lest they perish in the holocaust. After all, the firemen are the experts.

They heavily promote their special fire extinguishers as the only solution, even managing to get water outlawed for firefighting purposes! (Water wouldn't work on this kind of fire, they insist. Only the Fire Department's special extinguishers will suffice.)

Using a huge injection of taxpayer funds, the Fire Department gets their fire extinguishers built in record time, and they hard-sell them to everyone they possibly can. In the meantime, large swaths of the city burn to the ground. And due to the fire extinguishers' poor design and hasty construction, these devices turn out to be every bit as deadly as the fire, if not worse, for their damaging effects linger long after the fire has burned itself out.

But the firemen and their corporate cronies have secured their fortunes.

The bewildered, traumatized population can't figure out what happened, any more than the feckless politicians. The Fire Department emerges as the most powerful entity in the city. They resume their "research," fortified by their growing wealth and power.

After all, the next big conflagration is just around the corner.

Sound implausible? Think again. Because in the realm of "pandemic preparedness," the arsonists are running the Fire Department.

The Pandemic Preparedness Sweepstakes

Under the cover of vaccine development, there are dozens – perhaps hundreds – of biolabs around the world performing gain-of-function research on countless viruses and other infectious agents. The Wuhan Institute of Virology is the most infamous, but a great many of these labs are located in the United States, with at least 5 US labs manipulating H5N1 avian flu alone. This vast, shady industry of manufactured pathogenicity has infiltrated our government agencies, our military, and our universities, and of course, the pharmaceutical industry is thoroughly entwined in the whole enterprise.

Such "research" involves a multi-step process:

- obtaining grant funding – which also provides legal, intellectual, and ethical cover – for gain-of-function research, by promoting it as essential for "pandemic preparedness" and vaccine development

- obtaining pathogens (usually viruses) from nature that do not currently transmit to and among humans, but could be made to do so
- altering those pathogens genetically in the lab by adding, manipulating, or removing genetic material, to make them more transmissible and/or more deadly in humans
- speeding the evolution of these viruses by passaging them through mammals with immunological features similar to humans, as well as to human cell cultures
- publishing one's "achievements" of successfully enhancing the transmissibility and/or virulence of pathogens in the scientific literature, thereby securing continued grant support
- securing patents on key elements of the manufactured viruses to ensure royalties when and if a vaccine for the pathogen is developed
- waiting for (or perhaps causing) the escape of these pathogens into animal or human populations
- setting into motion the entire pandemic response/vaccine development juggernaut

This work violates the Biological Weapons Convention of 1975. But these labs persist in their work, under the false premise that their "research" is designed to protect the world's population from "rapidly emerging infectious diseases" by promoting vaccine development.

This is a lie.

The gain-of-function type research done in these labs genetically alters these animal viruses, empowering them to do easily and readily what they rarely do in nature: jump from species to species, spread readily among humans, and kill humans in significant numbers.

In essence, these researchers take viruses naturally found in animals, and which possess minimal-to-limited risk to humans, and alter them to make them highly transmissible and deadly to humans.

Why?

There is no legitimate rationale for this research. It's really this simple: if one truly wishes to protect the world's population from Godzilla, one does not deliberately and systematically create Godzilla in the lab.

Such research makes no sense when it comes to vaccine development, either. If one is concerned about existing pathogens, one should develop treatments that conquer those existing pathogens themselves.

Naturally occurring pathogens already have numerous targets for interventions – whether those interventions involve repurposing existing medications or developing new medications (including vaccines). We already have an armamentarium of existing medicines that are known to be effective against viruses. Sensible, ethical, indeed *sane* research would focus on strategies of targeting the existing chinks in the potential pathogens' armor, rather than creating new, lethal superbugs in the lab.

Unfortunately, there is much less money to be made and little power to be grabbed using the sane approach. Contrary to the alarmist claims, there simply aren't many naturally-occurring pandemics. And the enormous payoffs that Big Pharma and the investigators seek only come from patented, new, proprietary products – especially of the kind that can be put on a subscription model, like annual vaccines.

The Covid Pandemic as Dress Rehearsal

Of course, we have already seen the entire arsonists-running-the-fire-department scenario during Covid. A lab-developed, leaked pathogen prompted lockdowns. Patients who tested positive were told to stay home without treatment. Existing, established generic drug treatments with excellent safety profiles, such as hydroxychloroquine and ivermectin, were ruthlessly suppressed by the authorities – but *only* for use against the virus.

When patients became seriously ill, they were admitted to hospital and treated with proprietary medicines administered under directed protocols that later proved to be toxic to the patients, yet highly profitable to the drug manufacturers and patent holders. Meanwhile, the hospital systems were rewarded for their obedience with large bonuses for each Covid diagnosis made and each Covid death they presided over.

The proprietary "vaccines" were manufactured in record time (translation: far too quickly), and the most outrageous, coercive campaign to enforce medical treatment in history was unleashed, to compel the entire world to accept an experimental, rushed-to-market, misnamed "vaccine" based on the novel mRNA gene therapy platform. The results were devastating.

According to the CDC's own Vaccine Adverse Events Reporting System (VAERS), the Covid injections resulted in adverse events at a rate 117.6 times higher than the influenza vaccine.

As of May 30, 2024, more than 1.6 *million* adverse events have been reported to VAERS for the Covid-19 injections, as well as 38,559 deaths and 4,487 miscarriages. These numbers dwarf the VAERS reports for all other vaccines *combined*. By any measure, the Covid-19 mRNA injections were historically toxic and deadly interventions.

These data have accrued despite the fact that VAERS is a very laborious system in which to file a report and the fact that healthcare personnel who insisted on filing appropriate VAERS reports were harassed and sometimes even fired for doing so.

Furthermore, the compilation and publication of these data has been suppressed by the authorities and has only been revealed to the public by independent investigators. Additionally, there is a well-established underreporting error related to VAERS of at least one and perhaps two orders of magnitude.

Today, multiple of the Covid injections that were repeatedly touted by the authorities as "safe and effective" have been pulled from the market, including the Johnson & Johnson and Astra-Zeneca products. Ironically, the most dangerous ones remain.

Why? Because the survivors are mRNA products. The mRNA platform on which the "surviving" Covid injections are created presents a nearly unlimited potential for financial gain, as it provides an almost "plug and play" platform for gene therapies that can be marketed against future numerous infectious pathogens – as well as cancers and other diseases.

The Capture of Medicine and Academia

As mentioned above, hospital systems were drawn into this disreputable work by powerful financial incentives from both Big Pharma and captured government agencies. But hospitals are not the only formerly trusted institutions that have been drawn in.

Decades before Covid, many universities became implicated in bioweapons research, with highly profitable gain-of-function labs appearing at numerous of these prestigious institutions. These labs are funded by multiple problematic sources: government agencies such as Anthony Fauci's disgraced NIAID branch of the National Institutes of Health, Big Pharma, and private vaccine proponents/investors such as the ubiquitous Bill Gates.

Seminal work on the creation of SARS-CoV-2 – the virus that causes Covid – took place not in Wuhan but at the Ralph Baric Lab at the University of North Carolina at Chapel Hill. It's no stretch to say that since Covid-19, the world's most famous Tar Heel is no longer Michael Jordan – it's SARS-CoV-2.

At this writing, the same scenario is undergoing a terrifying reprise with the H5N1 influenza virus, commonly referred to as "avian influenza" or "Bird flu." As mentioned before, at least 5 labs in the United States alone are manipulating this virus, as well as multiple other labs abroad.

If the Bird flu does get out of the lab and become a pandemic, here are 2 key scientists (and their associated labs) to hold accountable:

Yoshihiro Kawaoka, PhD, of the Department of Pathobiological Sciences at the University of Wisconsin School of Veterinary Medicine, has been working on gain-of-function studies with avian influenza since 2006. He is funded by the Japanese government, as well as Daiichi Sankyo Pharmaceuticals, Fuji Corporation, and the Gates Foundation, among other sources.

Kawaoka is cofounder of the vaccine company FluGen. He holds 57 US patents, many of which are on Bird flu genetic sequences to be used for human avian influenza vaccinations.

Shockingly, the Kawaoka lab has been responsible for *two* known prior leaks of avian influenza. In the first, occurring in November 2013, a lab worker was stuck with a contaminated needle. While that fortunately did not lead to an outbreak, protocols were not followed both prior to *and after* this accident, leading to an NIH investigation that should have shut down the research entirely.

In the second accident, a lab worker in training lost a connection to his breathing tube and was exposed to air infected with respiratory droplets from ferrets infected with altered avian flu. Although this did not lead to infection, protocols were not properly followed yet again, and NIH was not appropriately notified of the accident.

As alarming as it is that such an accident-prone and

protocol-breaking lab is allowed to continue in any capacity, it is scandalous that Kawaoka's lab is now working with the same subclade (2.3.4.4b) of the H5N1 virus that has infected cattle in 12 states as well as three dairy workers.

One can only wonder what University of Wisconsin President Jay Rothman and the University of Wisconsin Board of Regents know (and do not know) about the Kawaoka lab's activities, and how they can justify sponsoring such potentially catastrophic "research" at the University they oversee.

Prof. R.A.M. (Ron) Fouchier, PhD, the Deputy Head of the Department of Viroscience at Erasmus University Medical Center in Rotterdam, the Netherlands, came to the forefront of avian influenza research in late 2011 when he successfully created a strain of the virus that could transmit in ferrets via aerosol respiratory droplets. This was a major step towards developing a virus that could transmit in humans, as the immune systems of ferrets and humans share considerable similarities.

This shockingly dangerous research earned Fouchier considerable criticism from even some of the most prominent pro-vaccine figures in medical research. The Foundation for Vaccine Research wrote a letter to the Obama White House in March 2013 condemning Fouchier's work, calling it "morally and ethically wrong," and stating the need to

> *consider the ethical issues raised by H5N1 gain-of-function research, especially experiments to increase the transmissibility of H5N1 viruses so they can be transmitted between humans as easily as the seasonal flu...[which could] cause a global pandemic of epic proportions that would dwarf the 1918 Spanish flu pandemic that killed over 50 million people.*

Notably, this letter was signed by multiple preeminent vaccine proponents such as the "Godfather of Vaccines" Dr. Stanley Plotkin, and famous vaccine advocate Dr. Paul Offit. Fouchier's gain-of-function work was so alarming that even the most zealous vaccine advocates took unusually strong action to halt it.

A temporary halt on gain-of-function research ensued in the United States but did not last. Fouchier has not heeded their warning, and no one at Erasmus University or elsewhere has stopped him. Fouchier has continued his gain-of-function work with different strains of avian influenza and has amassed 20 US patents, many of which are focused on his gain-of-function experiments.

The Current State of Bird Flu in the United States

H5N1 influenza, specifically subclade 2.3.4.4b, genome B3.13, is currently infecting over 90 herds of cattle in 12 different states. The first report of the virus in cattle was in March 2024. Reverse Transcriptase-PCR testing has returned positive for virus RNA in nasal secretions and the milk of cows. However, the cattle appear to recover from the virus with supportive treatment and the mortality rate is near zero. Active infection has not been reported in beef cattle.

There have been three cases of cow-to-human transmission of the virus, where infected humans were working with dairy equipment. The first two cases (Texas and Michigan) resulted in conjunctivitis (pink-eye) which cleared on its own in three days. In those cases, viral RNA was detected in eye secretions but not in nasal swabs. The third case (Michigan) resulted in a cough without fever, and eye discomfort with a watery discharge. Strangely, the complete genomic sequence of H5N1 for this case has yet to be released, despite the fact that the case was reported weeks ago. The other two cases appear to be consistent with the strain infecting cattle.

Several scientists have proposed that the current strain of H5N1 (subclade 2.3.4.4b, genome B3.13) circulating through cattle and to three humans in the US could have leaked from the USDA Southeast Poultry Research Laboratory (SEPRL) in Athens, Georgia. Hulscher et al. 2024 point out that the virus emerged in South Carolina extremely soon after identification in Newfoundland and Labrador. The timing doesn't make sense for natural spread because both identifications occurred in December 2021, meaning that the virus must have somehow transported nearly 1,700 miles in the same month – unless it was somehow leaked from the SEPRL facility. There is no publicly available sequence information for the Newfoundland identifications, which is most unfortunate.

However, gain-of-function research projects involving H5N1 commenced at SEPRL in April 2021 and continued through December 2021. No sequence information has been publicly released from these projects and USDA officials claim that such information does not exist. Very soon after the South Carolina identification, the virus spread to a bottlenose dolphin found off the coast of Florida and moved precipitously through wild birds and poultry in the Southeast and Midwest. The first identifications of genome B3.13 in poultry in the US were in chickens in Indiana (January 2022) and the first identification in dairy cattle was in March 2024, although the transfer to cattle may have been as early as December 2023.

Very recently, H5N1 virus isolated from cattle in the US was sent to the UK for further testing. A lab leak in this instance could lead to catastrophe given the rapid spread of the strain seen in the US.

The overriding concern is the accidental or deliberate release of a lab-developed H5N1 clade that is designed to transmit human to human. At this point, the accounts of individuals like Fouchier explaining the current Bird flu situation don't add up.

They propose that the virus crossed over from Europe to Newfoundland and infected an exhibition farm in December 2021. Then this supposedly spread – almost magically – to South Carolina (with two separate Genbank entries) in a wigeon and a blue winged teal on Dec. 30, 2021. There were no reports made between Newfoundland and South Carolina during this time which is at a minimum very curious.

The spread from South Carolina makes some sense from that point forward (i.e., to a bottlenose dolphin in Florida and later to poultry, starting in Indiana). The Athens, Georgia USDA lab SEPRL was doing work on H5N1 subclade 2.3.4.4b, genome B3.13 from April to December 2021 and this could have very well spread, via mallards or other wild birds, to the surrounding population.

The Return of "Fear Porn"

On Tuesday, June 4, 2024, Dr. Deborah Birx (the "Scarf Lady" of Covid-19 fame) stated to CNN that every cow in the US should be tested every week for Bird flu and that every worker should also be pool-tested. Birx made this absurdly impractical recommendation despite the facts that a) there is little to no mortality in cattle infected with Bird flu, b) the FDA has yet to change guidelines regarding consumption of raw or pasteurized milk, and c) such irresponsible use of the diagnostic tests would generate huge numbers of false positive results.

Even considering her performance during Covid, Birx must know that such willy-nilly testing will destroy the reliability of the PCR tests, the specificity of which is highly questionable to begin with. Making such impractical and counterproductive recommendations is quintessential "fear porn," and calling for such irresponsible testing appears to be a deliberate attempt at stoking panic, and perhaps even generating false-positive cases.

Another example of the "fear porn" approach to "pandemic preparedness" was recent claims by the World Health Organization (WHO) that a patient in Mexico died in April 2024 due to H5N2 influenza. Even setting aside the issue of relevance, as H5N2 is an entirely different strain of influenza than H5N1, the claim was false. The Mexican Health Secretary refuted the WHO's claim outright. The WHO later admitted their claim had been incorrect.

The WHO's initial, false claim was widely reported in the mainstream media. However, their retraction has been mostly buried, and the rare reports of the retraction that have been published have been deceptive. An ABC report by one Mary Kekatos acknowledging the retraction misleadingly claimed the WHO had stated the patient "died *with* the H5N2 strain of bird flu." Just one week earlier, Kekatos herself had written an article about the WHO's description of the case *titled* "1st fatal human case of bird flu subtype confirmed in Mexico: WHO." Of note, the WHO's initial report explicitly described "a confirmed fatal case of human infection with avian influenza A(H5N2) virus."

Even on the rare occasion when the mainstream media reports data refuting pandemic "fear porn," they appear unable or unwilling to do so with transparent honesty, and even such disingenuous admissions are buried in internet search results.

On a more rational note, Robert Redfield, MD, former director of the CDC during the first year of Covid-19, predicted in an interview with NewsNation that the next pandemic would be avian influenza. Redfield believes that this will be a lab-leaked version of Bird flu, stating that "the 'recipe' for making bird flu highly infectious to humans is already well established," recalling that gain-of-function research on the avian influenza virus was carried out in 2012, against his recommendations. In other words, he believes the arsonists are at it again.

Conclusion and Recommendations

If, in fact, any labs were to release weaponized H5N1 into the population, this would be the outright act of biological arson at least the equivalent of SARS-CoV-2's initial escape from the Wuhan lab, and given the precedent set by the Covid-19 disaster, even an accidental release would constitute an inexcusable act of mass murder.

The risk of this research is so great, the likelihood of leaks – be they accidental or deliberate – is so well-established and so high, and the stakes regarding human life are so potentially catastrophic, that gain-of-function research must be stopped altogether.

Dr. Jane Orient, MD, Executive Director of the American Association of Physicians and Surgeons, made the following common-sense recommendations in response to the continued H5N1 "fear porn" promoted by persons such as Deborah "Scarf Lady" Birx and the WHO, and the warnings of former CDC Director Robert Redfield:

> *We need to cancel the panic, monitor for, and isolate, sick animals. Same for humans. Research and use repurposed drugs for treatment. Disqualify the people responsible for the Covid debacle. Allow free discussion of opinions. Destroy the dangerous viral stocks and secure the labs, and be aware of who's paying for the research.*

Along those lines, here are our recommendations:

1. Citing the 1975 International Bioweapons Convention, immediately shut down ALL gain-of-function research in the US. As Dr. Orient states, this action must include securing the labs and destroying the viral stocks. Any resistance or interference with this should be subject to criminal punishment for Nuremberg Code violations.
2. Immediately call for the same to be done at all international labs (especially, but not limited to, Fouchier's lab in the Netherlands and the Wuhan Institute of Virology). Again, announce that any resistance at any level will be regarded as Nuremberg Code violations.
3. Pass prompt legislation that any and all intellectual property associated with completed gain-of-function research resides entirely in the Public Domain. Any vaccines or therapeutics developed from such research will be generic and non-proprietary.
4. Cease all present funding and outlaw any future funding for genetic manipulation of pathogens.
5. Common-sense approaches to respiratory viruses must be re-established, focusing on good hygiene, isolation of the sick (not the healthy), intelligent and free use of existing therapies, a *local-to-regional* (not global) approach to public health, and the complete removal of those with a record of failure and/or dishonesty during the Covid-19 period from the entire process, including the WHO.

Now is the time for citizens to loudly voice their concerns on this issue to elected officials *and* to other persons of authority who are responsible. For example, residents of Wisconsin should let Wisconsin Governor Tony Evers, Senators Ron Johnson and Tammy Baldwin, and their State Legislators know how they feel about the Kawaoka lab. Additionally, University of Wisconsin President Rothman and the Board of Regents should hear from any and all Badger alumni who do not want their alma mater to be the source of the next pandemic.

The State of Florida has outlawed gain-of-function research within its borders. Of course, the Federal Government should be pressured to act definitively to end such research at home and abroad, but other states should still follow Florida's lead on this issue. Every political entity, large and small, that prohibits gain-of-function research makes an important step in the right direction.

The arsonists must be fired from the Fire Department. The whole fear-driven and deception-based operation that is "pandemic preparedness" must be stopped. If it isn't, the Covid-19 experience will be converted from a once-in-a-lifetime trauma to a regularly recurring man-made disaster.

Open Letter to Students and Parents about Vaccines

Originally published on August 27, 2024 in *Brownstone Journal*, on August 28, 2024 by *No College Mandates*, and on September 5, 2024 by *Children's Health Defense*.

Dear Students and Parents:

Another school year is upon us.

For many people, including students, the Covid era seems to be over. That confusing, frightening, and traumatic time finally feels like a thing of the past.

However, the harms we suffered linger on. The educational losses and the mental health trauma suffered by millions of students are just two examples.

Also, the threat of future pandemics is constantly promoted by the World Health Organization (WHO) and other supposed experts. Their non-stop plague-of-the-month campaign has mutated from Covid, to "Disease X," to Bird flu, and now to Monkeypox (which, having failed to create sufficient fear last time, has been rebranded as "Mpox").

The threat of another pandemic does exist. But it does not come directly from nature, as the WHO and other alarmists claim.

Instead, the danger comes from human-manipulated viruses, as it was with the SARS-CoV-2 virus that got out of the Wuhan lab in late 2019. Countless gain-of-function labs, both in the United States and abroad, are playing dangerous games with nature, genetically altering viruses to make them more transmissible and more virulent.

These so-called scientists create *bioweapons*, and yet incredibly they defend this in the name of *vaccine research*! If this sounds too crazy to believe, too Sci-Fi to be true, consider:

Many, many laboratories start with viruses obtained from nature that infect animals and genetically change them so that, in their lab-altered form, they will infect humans and transmit among us. Such altered viruses are bioweapons. In the process, those doing this "research" obtain patents on the viruses and the alterations made to them. Why?

They do this so that when the bioweapons get into the human population, either by accident or intentionally, the same people and organizations that created the bioweapons can create the supposed remedy, the "countermeasure," in the form of a vaccine. They use government money to develop it, impose it on everyone else, and make gigantic profits in the process, since they hold the patent rights to the medical technology. This happened during Covid, and there is reason to believe it will happen again.

To put it another way, In the world of "Pandemic Preparedness," there are arsonists running the Fire Department.

Couple this with the already highly profitable traditional vaccine industry, and it is no surprise that the number of vaccines on the CDC vaccine schedules continues to increase.

Why does this concern students?

There are 2 reasons why this problem is so important to students:

- First, because governments and Pharma use school enrollment at nearly all levels of education as a key mechanism to push the ever-growing vaccine agenda.
- Second, because young people have the most to lose from these policies.

If you are a college student today, the number of childhood vaccines you have received may be 2 to 3 times the number your parents received. If you have a sibling or cousin in elementary school, the number may be even greater for them. Unless the current situation changes, these numbers will only continue to increase.

However, as you may know, children's health is significantly worse for your generation than it was for your parents, with dramatic increases in autism, allergies, and other chronic illnesses, among other problems. Autism in particular has increased alarmingly and in direct proportion to the increase in childhood vaccines.

However, governments and the medical establishment continue to push for more vaccines.

The Situation for College and Health Care Studies Students

During Covid, millions of college students were mandated to get the experimental Covid injections to be allowed to attend school. Colleges mandated the shots despite the facts that:

- These injections did not stop infection or transmission of the virus.
- Healthy young people had essentially zero risk of serious illness and death from Covid.

Since the Covid mRNA "vaccines" were given to the public, over 1.6 million adverse events and over 38,000 deaths related to these injections have been reported to the CDC's own Vaccine Adverse Events Reporting System (VAERS). Among these toxicities, increased rates of myocarditis – sometimes fatal – in young people, especially boys, have been demonstrated in recipients of the mRNA injections.

About 550 colleges in the US never required Covid vaccines for undergraduate students. However, some of those colleges would have mandated them if state law had not forbidden it, and still pressured their students to take the shots. (We encourage high school students and parents currently researching colleges to strongly consider those schools that never mandated the Covid vaccines, but to carefully read the notes in the above link, and consult with current students and their families.)

A great many students in health care fields are still required to get these Covid injections to continue their studies. And according to the student advocacy group No College Mandates, 20 American colleges and universities still mandate the Covid shots for their undergraduates.

There is zero legitimate medical indication for healthy children or young adults to get the Covid mRNA injections. None. Any institution insisting upon this, especially at this late date, demonstrates a total disregard for the physical health and well-being of its students.

As a physician and parent, I will say this regarding those last holdout colleges: I would not send a stray dog to any one of those schools to be housetrained, much less send a child to be educated.

The Situation for Elementary, Middle School, and High School Students

At present, five US states (California, New York, Maine, Connecticut, and West Virginia) allow no religious or conscience exemption from vaccination for children to attend public schools. In some of these states, significant pressure is put on physicians not to write medical exemptions either.

"Minor consent" – a child being legally able to consent to vaccinations or other treatments without parental consent – exists in multiple US states. For example, in California, minor consent is legal for some vaccines beginning at age 12, while in New York, there is no set lower age limit for a child to consent to Human Papilloma Virus (HPV) vaccination. Washington State is among the most extreme. The pro-vaccine organization Vaxteen describes Washington's "mature minor doctrine" as follows:

> *In Washington, minors of any age do not need their parent's consent to receive **all healthcare services, including vaccinations**. This is called a "mature minor doctrine" and essentially means that if you talk to your doctor/healthcare provider and they decide you are "mature enough" to make your own health care decisions, you can.*

Recently, the Vermont Supreme Court released a controversial decision in the case of *Politella v. Windham*, ruling that the government can vaccinate children without parental consent or legal recourse under some circumstances. As attorney John Klar explains:

> *The Vermont Supreme Court ruled that the Public Readiness and Emergency Preparedness (PREP) Act immunized school*

officials from "all state-law claims...as a matter of law." The Court did not address state or federal constitutional privacy protections or bodily autonomy, merely swallowing these paramount individual rights in a perverse, all-entrusting servitude to federal preemption by an omnipotent administrative state.

Across all levels of education, governments are using school enrollment to push the ever-growing vaccine agenda.

Safety Concerns about the mRNA "VINOs"

The Pfizer and Moderna Covid mRNA injections, while commonly called vaccines, are not true vaccines, but a type of mRNA-based gene therapy. In effect, they are Vaccines-In-Name-Only, or "VINOs." As pointed out by Rep. Thomas Massie (R-KY) and others, the CDC's definition of "vaccination" was altered during Covid to allow new types of drugs to be described as vaccines. (According to US Federal law, vaccine manufacturers have extremely protected legal status against lawsuits compared to makers of other drugs.)

Even with all the safety concerns surrounding traditional vaccines, these mRNA products may well prove to be much more toxic than traditional vaccines over time.

The mRNA technology platform has numerous mechanisms by which it can potentially be toxic to multiple organ systems. As outlined in detail by researchers examining the adverse reactions of the Covid mRNA VINOs, these systems include 1) cardiovascular, (2) neurological, (3) hematologic, (4) immunological, (5) oncological, and (6) reproductive.

The mRNA platform has not been adequately studied for the widespread use that it is now receiving. Traditional vaccines – which still have toxicities – often underwent up to 10 years of trials before being approved for widespread use. However, during the pandemic, the Covid mRNA shots were rushed into use in about *one year*, through the so-called "Operation Warp Speed" program and by being granted Emergency Use Authorizations (EUAs) from the Federal Government, thereby bypassing usual safety testing.

Because the mRNA VINOs were rushed into use so quickly, there have been no long-term safety studies done on them. None. How could there be long-term studies on a product brought to market in one year?

This is concerning for anyone who receives this type of injection, but it is especially so for young people. After all, young children and students may have to deal with the potential side effects and toxicities of these products for as long as a *century*.

Now that the Covid pandemic has receded, reasonable people might assume that Pharma and government agencies would slow down and properly study the mRNA platform and the VINOs they plan to produce using this technology. Unfortunately, the opposite has been the case.

On its own website, Moderna describes a pipeline of mRNA VINOs currently in development for Influenza, Respiratory Syncytial Virus (RSV), Cytomegalovirus (CMV), Epstein-Barr Virus (EBV), Human Immunodeficiency Virus (HIV), Norovirus, Lyme disease, Zika virus, Nipah virus, Monkeypox, and others.

Considering the aggressive approach to the use of the mRNA platform by companies like Pfizer and Moderna, it is possible and perhaps likely that in the near future, traditional antigen-based vaccines may be mostly or even entirely replaced by mRNA-based "VINOs."

Many physicians and scientists who are not industry or Government insiders believe the use of the mRNA platform as a whole should be put completely on hold while the full scope of its toxicities, both short and long-term, are properly studied. However, the opposite is currently happening.

Summary

At the start of a new school year, who wants to be confronted with these scary topics? It is only natural that we should want things to return to the way they were before Covid. But these issues are both real and important, and every one of us must be aware of them and must face them, for the sake of our own health and well-being.

This letter cannot and should not be considered personal medical advice. Unlike the thousands of colleges, employers, and others who mandated the Covid shots, this letter is not here to tell you what to do. On the contrary.

Instead, this letter is intended to serve two purposes:

First, to inform and emphasize to students and parents that the situation with both traditional vaccines and the newer mRNA VINOs is problematic at best, and very harmful at worst. Even now, when Covid seems behind us, the situation is becoming more concerning, especially for young people.

Second, to remind all young people of their fundamental human right to determine for themselves what is and is not done to their body. In the field of medical ethics, this is called *autonomy*. It is the first and foremost pillar of medical ethics. Much of what happened in medicine during Covid was unethical and immoral. But that does not change the fact that doctors, nurses, health care systems, schools, universities, and government agencies have an ethical and moral responsibility to respect your bodily autonomy. Don't ever let anyone convince you otherwise, and don't trust anyone who tries to deny you autonomy over your own body.

In conclusion, to students and parents: please consider the following.

1. **Do your own research before you accept any medical treatment – be it a vaccine or anything else.** Do not accept any medical therapy simply out of convenience, or because someone else "mandates" it. Do not "take one for the team." Accept nothing based only on the assurances of others – not those of your school, your government, your doctor, and certainly not Pharma. Do your own research, maintain a healthy skepticism, and seek honest and impartial information about things you don't understand. Do not allow anyone to do anything to you that you are not confident is both safe and in your best interest.
2. **Learn to say "no."** There are many times in life when it is better in the long run to say "no," but easier in the moment to say "yes." Such careless compliance can often bring you to grief. Sadly, this is true in the world of medicine, as it is elsewhere in life.
3. **Learn what your options are and pursue them.** If you're a high schooler looking into colleges now, consider those schools that never mandated Covid vaccines. If you're a healthcare student still facing mandates, learn about exemptions and pursue one if you wish. If you're a parent of young children in an area where you feel there is excessive state overreach in public schools, consider alternative schools, homeschooling, or moving. Options still exist for students both to get an education and protect their health and autonomy, but you must learn what the options are and actively pursue them.

4. **Speak out.** Until the "Pandemic Industrial Complex" is stopped and the mRNA and vaccine pipelines are brought under proper control, the bodies of our young people will continue to be subjected to more novel pharmaceutical products of unknown toxicity. Is this the world you want to live in? For college students: does your school conduct gain-of-function research? Students at Wisconsin, North Carolina, Boston University, and numerous other institutions should consider taking action to get their alma maters to divest from such activity. Consider starting a student organization to advocate and protect your rights. Two good existing organizations you may wish to contact are Students Against Mandates and No College Mandates.
5. **Most of all, protect yourselves.** Your elders have mostly failed you in this matter. You have many years of life ahead of you. It is up to you to safeguard your own health.

May every student have a happy, successful,
and *healthy* school year.

Could Bird Flu Be the October Surprise?

Originally published on September 21, 2024 in *Brownstone Journal* and on September 24, 2024 in *The Defender of Children's Health Defense* as "Bird Flu or Monkeypox – Which Will they Weaponize Next?"

Bird flu was the hot topic in pandemic fear-mongering until very recently. Just a few months ago, former CDC director Robert Redfield publicly described Bird flu (also known as H5N1 Influenza A or Avian Influenza virus) as the likely next pandemic – predicting a laboratory-leaked virus as the cause. Meanwhile, Deborah Birx, aka the "Scarf Lady" of Covid infamy, was making the TV news, promoting an unrealistic and excessive program of testing farm animals and humans for Bird flu.

At present, Bird flu seems to have been put on the back burner by the authorities. Monkeypox has since taken center stage, with the World Health Organization declaring a state of emergency over that virus. Furthermore, the "experts" have trotted out numerous other viruses with which to terrify the public. Examples include West Nile virus – who no less than Anthony Fauci himself supposedly contracted – and even the exotic "Sloth virus" (also known as Oropouche virus).

The first step in dealing with these continual reports of horrific pathogens is recognizing the vital importance of living in knowledge rather than in fear. "Fear porn" is a real psychological weapon and one that is being used against us on a daily basis. As we painfully learned during Covid, a terrified population is easily manipulated, controlled, and exploited. As free citizens, we must remain mindful and knowledgeable, rather than fearful, about the flood of information and propaganda that is hurled at us.

Regarding Bird flu, we should remain mindful of the following. In its current iteration, Bird flu has caused no widespread human illness, no human deaths, and sporadic outbreaks in farm animal populations. However, there is much evidence that Bird flu could be used as a bioweapon.

Furthermore, it could also be applied to disrupt the November 5 US Presidential election.

Here are 3 reasons why Bird flu may still be weaponized to alter the election:

- Multiple bio labs in the United States and abroad – such as the lab run by Yoshihiro Kawaoka, PhD at the University of Wisconsin – perform alarming gain-of-function research on the H5N1 virus, making variants of the virus that are much more dangerous to humans than variants that occur in nature. These labs have had leaks with alarming frequency. The current strains of Bird flu in the US show strong genetic evidence of having originated in a laboratory. A laboratory leak of a new strain of the virus, manipulated to be highly transmissible and/or pathogenic in humans, remains a real possibility.

- The "International Bird Flu Summit" will be held on October 2-4, 2024 at the Hilton Fairfax in Fairfax, VA – just outside Washington, DC – exactly one month prior to the election. Listed topics include "Command, Control and Management," "Emergency Response Management," and "Surveillance and Data Management." If this sounds eerily reminiscent to you of the Covid lockdowns – which were also closely preceded by government-based planning exercises – your memory serves you well.
- The infrastructure is already in place for a "pandemic" of Bird flu, much more than it is for other potential pathogens. Already, widespread testing of farms is underway. The development of Bird flu vaccines has increased dramatically. The FDA has already approved vaccines made by Sanofi, GSK subsidiary ID Biomedical Corporation of Quebec, and CSL Seqirus, while Moderna recently received a $176 million government grant for its mRNA-based Bird flu injection, which is in development.

In the bigger picture, a number of viruses could potentially be employed as an "October Surprise" to disrupt the election. Bird flu appears to be a leading candidate (pun intended), but it is not the only one.

We, as citizens, must remain vigilant to this threat to our electoral process. We should contact our local and state officials now, *before* anything is attempted, and express our absolute insistence on fair, legal, and regular elections. We should share this information widely with others so that all are aware of what might be attempted. Over the longer term, we must work to end gain-of-function research.

With Covid, we experienced first-hand what can be done to our civil rights and to our Constitutionally guaranteed electoral and governmental processes when a fear-driven, emergency-based takeover of society occurs. As free citizens, we must never allow this to happen again. From now on, we must live in knowledge, not in fear.

Monkeypox: Evidence of the "Pandemic Preparedness" Lie

Written with Brian Hooker, PhD, and Heather Ray.
Originally published October 14, 2024 by *Brownstone Institute* and on October 16, 2024 in The Defender *of Children's Health Defense.*

"Pandemic Preparedness," and the gain-of-function research that underlies it, operates under a grand deception, a big lie. The Biological Weapons Convention, which every major nation has signed, "prohibits the development, production, acquisition, transfer, stockpiling and use of biological and toxin weapons." As a result, gain-of-function research – the process of taking viruses and other pathogens found in nature and making them *more* transmissible and dangerous in humans – must be justified by defining it as something other than what it really is – namely, the creation of biological weapons and countermeasures for those weapons.

The grand deception – the big lie – used to justify gain-of-function research goes something like this: "We need to alter pathogens in the lab to anticipate the mutations that just might occur in nature, and to promote the production of vaccines to protect humanity from these theoretical superbugs."

In truth, there is no legitimate reason to create superbugs in the laboratory. One does not save Tokyo by creating Godzilla. Unfortunately, science can be both complicated and confusing, especially when the "experts" are intentionally untruthful. This grand deception has therefore worked for decades, and a gigantic, profitable, and frankly terrifying pandemic preparedness industry involving governments, non-governmental organizations, Big Pharma, and universities has grown as a result.

In order to expose and discredit a big lie that has persisted for such a long time, sometimes a "smoking gun" is needed – that is, a piece of clear and obvious evidence that the long-held premise is false. In the case of the big lie surrounding gain-of-function research and the pandemic preparedness industry, monkeypox serves the role of smoking gun.

Monkeypox virus is back in the news in 2024, as one of the pandemic industrial complex's leading candidates for the so-called "Disease X" about which the World Health Organization has been sounding its relentless alarm. (Of course, this is the second time monkeypox has been trotted out in recent years, after the 2022 monkeypox fear porn campaign in the United States that ultimately fizzled out.)

Once one gains a thorough understanding of both the monkeypox virus's peculiar history in the US, as well as the natural characteristics of the virus, one can easily see through the grand deception – the big lie – that is used to justify gain-of-function research and the entire "pandemic preparedness" industry.

Monkeypox Comes to America

In 2003, through exotic pet importation, 35 people in six US states were confirmed to have been infected with the clade II type of the monkeypox virus. The humans contracted the disease from infected prairie dogs, kept as pets, that had themselves been exposed to either contaminated imported animals or other individuals infected with the virus. All human cases made a full recovery without lasting effects.

This outbreak was an odd, self-limited, and entirely incidental occurrence of a rare and essentially non-lethal virus finding its way to the US by specific and preventable circumstances. In a world of sensible and ethical public health practices, this event should have prompted a reasonable, proportionate response, such as increased precautions regarding the exotic animal trade.

Instead, this incident opened the floodgates to dangerous research by scientists who sought to identify a strain of monkeypox that could easily be passed to humans by way of aerosol transmission.

In 2009, Christina Hutson and her team at the CDC collaborated with Jorge Osorio at the University of Wisconsin to investigate the transmissibility of monkeypox. Again, in 2012, Hutson teamed with other universities to test and compare the transmissibility of the monkeypox virus in rodents, ultimately determining in those experiments that "transmission of viruses from each of the MPXV clade was minimal via respiratory transmission."

Again, in a sensible and ethical world, these findings might have shut the door on ill-advised research on monkeypox. As we shall see, that was not the case.

Monkeypox: A Lumbering Giant of a Virus

The monkeypox virus itself is a strange candidate indeed to try to manipulate in the manner Hutson and Osorio sought. Unlike small, simple, rapidly mutating RNA respiratory viruses like

Influenza viruses or coronaviruses, monkeypox is, in the virus world, a slow-moving, lumbering giant.

The most 'successful' bioweapon in human history is the SARS CoV-2 coronavirus that causes Covid. It encodes only 29 proteins in its single-stranded, RNA genome, which is correspondingly small – slightly less than 30,000 bases in length. With its genetic simplicity and its single-stranded RNA genome, it mutates very rapidly. The virus itself is small as well – it is only about 100 nanometers in diameter and weighs about 1 femtogram (or 0.000000000000001 gram).

As one might expect, this virus is readily transmitted through the airborne route.

Monkeypox virus, by contrast, is one of the largest and most complex viruses in existence. It can be up to 450 nm long and 260 nm wide, and its double-stranded DNA genome has nearly 200,000 base pairs. With this lengthy, complex genome, encoded in more stable, double-stranded DNA, it mutates slowly. This large virus – a giant, by viral standards – does not transmit by the aerosol route. Rather, it is transmitted by close contact, including sexual intercourse (as became well known during the 2022 monkeypox scare), as well as the hunting, slaughtering, and eating of bushmeat.

Consider also that naturally occurring monkeypox is much less deadly to humans than the pandemic planners and fear pornographers typically advertise. The WHO has since reported on the international monkeypox outbreak that occurred in 2022. As of January 2023, the total number of confirmed cases was 84,716, with 80 total deaths. Thus, the case fatality rate during that outbreak was less than one death in every thousand cases, *100 times less* than the frequently-cited case-fatality rate of 10%.

Strictly speaking, the frequently cited 10% case-fatality

rate refers only to the more virulent clade I of monkeypox. However, many authorities have picked up the bad habit of bandying about the 10% figure indiscriminately of clade. Furthermore, even with clade I, this rate appears to be a significant exaggeration.

For example, on its webpage on endemic clade I Monkeypox in the Democratic Republic of the Congo, the CDC states that "Since January 1, 2024, the Democratic Republic of the Congo (DRC) has reported more than 31,000 suspect mpox cases and nearly 1,000 deaths." These numbers result in a case fatality rate of around 3%.

However, on October 13, 2024, the WHO released an updated "situation report" on Monkeypox that demonstrates current case-fatality rates for *confirmed* cases of monkeypox to be much lower still. According to this report, from January 1 to August 31, 2024 there have been 106,310 confirmed cases worldwide with only 234 confirmed deaths. This corresponds with a case-fatality rate of 0.0022 – only 0.22%, or 1 death in every 454 cases.

Even in the Democratic Republic of the Congo (DRC), where the reportedly more deadly Clade I is endemic, the WHO reports 6,169 confirmed cases of monkeypox so far in 2024 with just 25 deaths, resulting in a case-fatality rate of 0.4%. This is nearly an order of magnitude less than the case-fatality rate for 'suspected' cases.

Finally, the WHO reports that there have been zero confirmed monkeypox deaths out of 2,243 confirmed cases in Africa (the majority of these in the DRC) during the last 6-week reporting period. *Zero deaths.*

Put simply, the WHO's own official reports directly contradict the monkeypox fear porn that is being promoted worldwide, and calls into serious question the data on 'suspected' cases.

There are numerous other threats to human health that are more worthy of time, funding, and effort. For example, in the Democratic Republic of the Congo, where monkeypox is endemic, about *eighty times* more people die of malaria than of monkeypox. Malaria is both preventable and curable with proper diagnosis and access to inexpensive medications. This tragic death toll from malaria illustrates how common, deadly, but relatively unprofitable diseases are neglected by supposedly philanthropic entities such as the WHO.

Instead, they heavily promote the grand deception of pandemic preparedness and gain-of-function research.

Given the monkeypox virus's sheer size, complexity, low rate of mutation, relatively stable DNA genome, and instability when exposed to oxygen, the likelihood of it ever naturally mutating into an airborne pathogen is remote. There is simply no legitimate reason to monkey with its genome in the lab (pun intended).

Add to the mix its limited transmissibility and low mortality (especially for clade II), and any honest and competent scientist truly seeking to serve humanity would recognize that naturally occurring monkeypox is a relatively low public health priority and a marginal-at-best vaccine candidate – especially for the world population at large.

But Anthony Fauci and his cronies at NIAID saw things differently.

Fauci and Friends, at It Again

In 2015, Anthony Fauci's National Institute of Allergy and Infectious Disease (NIAID) covertly approved a dangerous gain-of-function experiment that would genetically manipulate the monkeypox virus to create a more virulent and transmissible pathogen that would potentially pose a grave threat to humans.

Instead of raising the alarm about this proposal to create a deadly hybrid monkeypox virus, the Department of Health and Human Services (HHS), the National Institutes of Health (NIH), and NIAID itself deceptively hid the project's approval from the oversight of the House Committee on Energy and Commerce, by burying funding for the experiment in an alternate grant.

The project was proposed by Dr. Bernard Moss, a long-time friend and colleague of Fauci at NIAID. Moss, who has accumulated multiple US patents related to monkeypox, intended to insert virulence genes from the more severe form of monkeypox, clade I (Congo Basin clade), in the "backbone" of the more transmissible monkeypox virus, clade II (West Africa clade). This project would create a much more dangerous version of monkeypox with the virulence of clade I and the transmissibility of clade II. This chimeric form of monkeypox would not originate in nature, as different clades of DNA viruses do not naturally transpose genes.

It is unknown whether this ill-advised, highly dangerous, and deceitfully approved project was completed. Fauci and Moss's sleight-of-hand was discovered in 2022, prompting a seven-month Congressional investigation. The House Committee Report (page 6) states that "HHS, the NIH, and NIAID continue to insist the GOFROC (gain-of-function research of concern) experiment transferring material from clade I to clade II was never conducted, despite being approved for a period of over 8 years. However, HHS has repeatedly refused to produce any documents that corroborate this claim."

Is a weaponized form of monkeypox in existence? If so, Fauci, Moss, and friends aren't telling.

What is known is that there was no legitimate reason to conduct such experiments, and that those involved knew this, as they hid the project from their overseers. The only logical assumption about the intent of the research is that it was to create a weaponized version of monkeypox.

The House Committee's conclusions on Fauci's NIAID as a whole are damning:

> *The primary conclusion drawn at this point in the investigation is that NIAID cannot be trusted to oversee its own research of pathogens responsibly. It cannot be trusted to determine whether an experiment on a potential pandemic pathogen or enhanced potential pandemic pathogen poses unacceptable biosafety risk or a serious public health threat. Lastly, NIAID cannot be trusted to honestly communicate with Congress and the public about controversial GOFROC experiments. (page 8)*

NIAID couldn't be trusted about Covid.

They cannot be trusted about monkeypox, either.

According to the House Committee on Energy and Commerce, they cannot be trusted, period.

To summarize: in nature, monkeypox disease is a relatively rare, usually mild viral illness transmitted through behaviorally modifiable forms of close contact such as sexual intercourse and the hunting and eating of bushmeat. The infectious agent is a very large, complex DNA virus that transmits poorly from person to person and is much less prone to mutation than numerous other viruses.

Once one realizes all this, it becomes frankly preposterous to attempt to justify gain-of-function research on such a pathogen for any legitimate purpose. The only plausible reason to do such research on monkeypox is to create a bioweapon – a weaponized virus – and to also create and profit from its countermeasure – a proprietary vaccine.

Pandemic preparedness is a grand deception, a big lie. The monkeypox madness demonstrates this, as compellingly as a smoking gun at a murder scene. We must put an end to all gain-of-function research and to the bogus pandemic preparedness excuse for illegal bioweapons research.

Six Simple Steps to Pharma Reform

Originally published on November 20, 2024 by *Brownstone Institute*, on November 21, 2024 by *The Defender of Children's Health Defense*, and on November 21, 2024 by *Science, Public Health Policy, and the Law*.

The recent United States elections may have finally produced an administration that is willing – even eager – to reform the Big Pharma juggernaut that has thoroughly dominated life in the United States since Covid. But how might we achieve meaningful, definitive Pharma reform?

Simple.

Before we continue, please allow me to highlight the difference between "simple" and "easy." Just because something is simple doesn't make it easy. Lifting a 10-ton weight is no more complicated than lifting a 10-pound weight. But it's a lot harder to do.

The task of reforming Big Pharma will not be easy. Talk about a heavy lift! Consider that before the 2020 election, the pharmaceutical industry donated funds to 72 senators and 302 members of the House of Representatives. Pfizer alone contributed to 228 lawmakers. At this moment, Big Pharma may be down, but it's not out. The industry has too much power, money, and influence to be brought under control without a major struggle.

While not easy, should the political will be mustered, the process of breaking the stranglehold Big Pharma has on us would be surprisingly simple. Six changes in Federal law – four repeals of existing law, and two new pieces of legislation – would go a long way toward reining in and even reforming Big Pharma.

From the 1970s onward, US Federal policy consistently trended toward the empowerment and enrichment of the pharmaceutical industry. Since 1980, a series of Federal laws were enacted that created perverse incentives and promoted the rapacious behavior that has characterized Big Pharma over the past several decades, climaxing with the pandemic totalitarianism of the Covid era.

Four of the most problematic of these laws are ripe for repeal. Doing so would constitute vital steps toward reining in Big Pharma. The two other steps proposed here would require new legislation, but fairly simple legislation at that.

The six simple steps are:

- Repeal the 1980 Bayh-Dole Act
- Repeal the 1986 National Childhood Vaccine Injury Act
- Repeal the 2004 Project Bioshield Act
- Repeal the 2005 PREP Act
- Outlaw Direct-to-Consumer Pharmaceutical Advertising
- Encode Medical Freedom into Federal Law

Repeal the 1980 Bayh-Dole Act
The Patent and Trademark Law Amendments Act (Public Law 96-517), better known as the Bayh-Dole Act, was signed into law by Jimmy Carter in 1980.

The Bayh-Dole Act made 2 major changes: it allowed private entities (such as universities and small businesses) to routinely keep ownership and patent rights to inventions made during government-funded research. It also allowed Federal agencies to grant exclusive licenses for use of Federally-owned patents and intellectual property.

The Bayh-Dole Act was intended to encourage innovation within government research. As researchers could now profit directly from their work, it was thought they would make better use of taxpayer support. However, as economist Toby Rogers has argued, this ill-conceived law had the opposite effect.

The ability for government contracted workers to patent their discoveries created a disincentive to share them with other researchers, who might beat them to market. Close guarding of intellectual property and lack of open collaboration had a chilling effect on rapid innovation – hardly what taxpayers would have wanted from their investments.

More importantly, endowing Federal agencies such as the NIH with the power to effectively pick "winners and losers" with whom Federal intellectual property would be granted for commercial use, created a tremendous potential for corruption within these agencies.

The Act did contain a provision for "march-in-rights," whereby the relevant government agency (such as the NIH) could step in and allow other entities use of the intellectual property if the original patent-holder failed to meet specific requirements to make proper use of them for the public good. However, according to the US Chamber of Commerce, in 44 years since the Act was made law, march-in-rights have never been successfully invoked, despite numerous attempts.

The Bayh-Dole Act itself, coupled with the refusal of agencies such as the NIH to ever invoke march-in-rights, has been frequently implicated in the massive price-gouging problems in US pharmaceuticals. In one remarkable exchange in 2016 between Senator Dick Durbin and then NIH Director Francis Collins, Durbin refuted Collins' prevaricating defense of never invoking march-in-rights, stating:

> ...if you cannot find one egregious example where you could apply this [march-in-rights], I would be surprised. And applying it even in one, sends at least the message to the pharmaceutical companies, that patients need to have access to drugs that were developed with taxpayer's expenses and the research that went into it. I think that doing nothing sends the opposite message, that it's fair game, open season, for whatever price increases they wish.

By allowing the NIH authority to assign publicly funded intellectual property rights *and* statutory power to protect exclusive use of them, the Bayh-Dole Act opened the door widely for massive corruption between industry and regulators and greatly enabled the extreme degree of agency capture now present at the NIH and other Federal Agencies.

Bayh-Dole has been a failure. It should be repealed and replaced.

Repeal the 1986 National Childhood Vaccine Injury Act

The toxicity of vaccines was so well-established even decades ago, that a Federal law – the National Childhood Vaccine Injury Act (NCVIA) of 1986 (42 U.S.C. §§ 300aa-1 to 300aa-34) was passed to specifically exempt vaccine manufacturers from product liability, based on the legal principle that vaccines are "unavoidably unsafe" products.

Since Ronald Reagan signed the 1986 NCVIA Act protecting vaccine manufacturers from liability, there has been a dramatic increase in the number of vaccines on the market, as well as the number of vaccines added to the CDC vaccine schedules, with the number of vaccines on the CDC Child and Adolescent schedule rising from 7 in 1986 to 21 in 2023.

Furthermore, this special protection afforded to vaccines has prompted Big Pharma to attempt to sneak other types of therapeutics under the "vaccine" designation to provide them with blanket liability they would not otherwise enjoy.

For example, the Pfizer and Moderna Covid mRNA injections, while commonly called vaccines, are not true vaccines, but rather a type of mRNA-based gene therapy. In effect, they are what I refer to as Vaccines-In-Name-Only, or "VINOs." As pointed out by Rep. Thomas Massie (R-KY) and others, the CDC's definition of "vaccination" was altered during Covid to allow new types of drugs to be labeled as vaccines.

We have now reached the previously unimaginable state where Big Pharma is touting potential "vaccines" for cancer. As the National Cancer Institute admits on its website, these are actually immunotherapies. The purpose of employing this misleading nomenclature is clear: to slide even more therapies under the tort-protected "vaccine" umbrella.

The bloom is off the rose for vaccines. The alarming toxicity of the Covid vaccines caused a worldwide reexamination of this entire class of medicines. Multiple Covid vaccines, including the Johnson & Johnson and AstraZeneca products, once brazenly touted as "safe and effective," have now been pulled from the market. And the literally millions of VAERS reports implicating the mRNA Covid products have not gone away.

The National Childhood Vaccine Injury Act (NCVIA) of 1986

should be repealed, returning vaccines to the same tort liability status as other drugs.

Repeal the Project Bioshield Act of 2004

The Project Bioshield Act, signed into law by George W. Bush in 2004, introduced the Emergency Use Authorization avenue for pharmaceutical products to be brought to market. Among other things, this law empowered the FDA to authorize unapproved products for emergency use, in the event of a public health emergency as declared by the Department of Health and Human Services (HHS).

By its very design, this law is ripe for abuse. It places immense power in the hands of the unelected Director of HHS, who can declare an emergency activating the law, and who simultaneously oversees the FDA.

This power was egregiously misused during Covid. Shockingly, the FDA issued nearly 400 EUAs related to Covid for pharmaceutical and medical products, the Covid "vaccines" being only the best known. The FDA even went so far as to grant "umbrella" EUAs for entire *categories* of Covid products such as test kits, often without reviewing specific products at all. The immense amounts of fraud related to test kits and other Covid-era medical products should come as no surprise.

With regard to Covid-related pharmaceuticals, to this day EUAs continue to be misused to the benefit of Big Pharma and to the detriment of citizens. For example, when the FDA announced the "new" formulations of the Covid boosters for 2024-25, they still released these new products under *Emergency Use Authorization*. In other words, a full *four-and-one-half years* after the start of the Covid pandemic, these products are still rushed to market after ludicrously inadequate safety and efficacy trials, based on a purported "emergency" now approaching a half decade in length.

The 2004 Project Bioshield Act should be repealed and the EUA designation it created should be eliminated.

Repeal the PREP Act of 2005

The NCVIA already provided vaccine manufacturers with a blanket tort liability shield beyond the wildest dreams of other industries, but apparently that was not enough. In 2005, at the height of the "War on Terror," George W. Bush signed the Public Readiness and Emergency Preparedness Act (42 U.S.C. § 247d-6d), better known as the PREP Act.

The PREP Act, which was heavily lobbied for by vaccine manufacturers, provides an unprecedented level of blanket tort liability to Big Pharma and other medical-related industries in the event of declared bioterrorism events, pandemics, and other emergencies. Again, tremendous power is placed in the hands of the Director of HHS, who has broad discretion to declare such an emergency.

The PREP Act was controversial from the outset – any act that can spark vigorous, simultaneous opposition from both Phyllis Schlafly's conservative Eagle Forum and Ralph Nader's left-wing Public Citizen for its unconstitutional nature is surely pushing the envelope.

In effect, the PREP Act has allowed Big Pharma and its captured regulatory friends to completely circumvent routine FDA standards for safety and efficacy under the guise of an emergency, which as noted above, can conveniently last half a decade or more.

Furthermore, in the aftermath of Covid, the PREP Act has been broadly invoked in the legal defense of countless defendants now sued for the excesses, harms, and violations of human rights perpetrated at all levels of government and society. It will take decades in the courts to sort out where the PREP Act's broad protections begin and end.

This is both absurd and insane. At its inception, the PREP Act was broadly recognized as one of the most overreaching and unconstitutional Federal laws in modern times. The Covid era has tragically revealed the PREP Act to be a murderous failure. The PREP Act must be repealed.

During Covid, government at nearly every level used the specter of a pandemic to blatantly suspend, deny, and even attempt to permanently eliminate numerous fundamental civil rights that are clearly encoded in the Constitution. Furthermore, the well-established and time-honored pillars of Medical Ethics were dismissed wholesale in the name of public safety.

In addition to repealing the deeply flawed laws discussed above, two pieces of straightforward legislation are needed to limit Big Pharma's undue influence on society.

Outlaw Direct-to-Consumer Pharmaceutical Advertising

The United States is one of only 2 countries in the world that allows direct-to-consumer advertising of pharmaceuticals. The scale of this advertising is monumental. Total Pharma advertising spending topped $6.58 billion in 2020. The dangers of this are multiple.

First, as we can all see by turning on the television, Big Pharma abuses this privilege by aggressively hawking almost any product it feels it can profit from. The "pill for every ill" mindset shifts into hyperdrive on TV, with an expensive, proprietary, pharmacological cure for everything from your morbid obesity to your "bent carrot."

Direct-to-consumer television advertisements heavily target

the elderly. This is an important component of Big Pharma's push to promote the Covid and RSV vaccines as routine shots, piggybacking on the wide acceptance of influenza vaccines. Not content to profit off the traditional fall flu vaccine, Big Pharma seeks to create a subscription model for a bevy of seasonal shots against numerous, generally mild, viral respiratory infections.

Even more importantly, direct-to-consumer advertising provides Big Pharma with a legal way to capture media. Pharma was the second-largest television advertising industry in 2021, spending $5.6 billion on TV ads. No legacy media outlet dares to speak out against the interests of entities providing that level of funding. This muzzles dissenting voices and eliminates open discussion about safety issues in mainstream media.

In short, through direct-to-consumer advertising, Big Pharma has bought the media's silence.

A free society requires freedom of the press and media. The Covid era has demonstrated that direct-to-consumer pharmaceutical advertising stifles freedom of the press and media to a dangerous and unacceptable degree.

Somehow, the rest of the world has managed to survive without direct-to-consumer pharmaceutical advertising. In fact, many countries do better with respect to health measures than the Pharma-ad-riddled USA. In 2019, just before Covid, the United States ranked only 35th in terms of overall health in the Bloomberg National Health Rankings. Meanwhile, the United States pays more for its middling health rankings than any other nation on Earth.

Encode Medical Freedom into American law

The Founding Fathers would be scandalized to find that the United States needs explicit laws stating that the Bill of Rights is *not* null and void in the event of a "pandemic," (or during other emergencies, for that matter), but here we are.

The Founders were well acquainted with episodic infectious disease. In fact, they faced epidemics at a level we cannot imagine. George Washington survived smallpox. Thomas Jefferson lost a child to whooping cough. Dr. Benjamin Rush, signer of the Declaration of Independence and surgeon general of the Continental Army, promoted inoculation of the troops against smallpox.

Despite those experiences, the Founders inserted no health-emergency-based escape clauses in the Constitution permitting government to deny citizens the inalienable rights protected therein.

As I have written previously, the excesses of the Covid era have sparked a movement toward encoding "medical freedom" into law, to protect our civil rights against medical and public health overreach. (To be fully effective, this may need to be expanded to include any declared emergency – e.g. "climate" emergencies – although that is beyond the scope of this essay.)

Given the excesses of the Covid era, many of which have now been demonstrated to have been preplanned and deliberate, and given rapid technological advancement of both medicine and surveillance, it is advisable to encode into law assertions regarding medical freedom. While the exact wording may vary, the 2 key points of focus would be explicitly protecting bodily autonomy and limiting the power of public health declarations. Here are two examples:

- *Citizens shall not be deprived of any rights protected in the US Constitution, or of their ability to fully participate in society, on the basis of their acceptance or refusal of any medical treatment(s) or procedure(s).*

- *Citizens shall not be deprived of any rights protected in the US Constitution, or of their ability to fully participate in society, on the basis of a medical or public health emergency.*

Encoding such statements into law would accomplish two goals. First, it would substantially rein in the power-seeking element of the public health industry that became such a menace to human freedom during Covid, and which incidentally is tightly entwined with Big Pharma. Second, it would significantly thwart the efforts of Big Pharma to push their wares through a herd-based and mandate-driven approach.

Should someone oppose such explicit statements of our God-given rights, on the basis of "But what if there is another pandemic?", I would reply as follows: Only once in human history did the world lock itself down due to a disease. It turned out to have been done mostly under false pretenses, and it turned out to be a deadly and disastrous mistake. We are not doing that again.

Conclusion

Big Pharma is a Leviathan, in both the biblical and Hobbesian senses of the word. To truly control it, other measures will surely be necessary. Other needful actions are beyond the scope of this article. Some of these may be very complicated. For example, it is imperative that the gain-of-function bioweapons research be halted. However, this is a worldwide issue, so outlawing it in the US alone will not solve the problem.

However, these six simple steps are an important start. Members of the incoming administration have already spoken about some of them. Success breeds success, and successfully implementing these solutions will help free ourselves from the tentacles of the monstrosity that Big Pharma has become.

The Pandemic Planners Come for Hoof and Hen...and Us Again

Originally published on January 4, 2025, by *Brownstone Institute*, and on January 7, 2025 by *Daily Clout*.

On December 31, 2024, the world received a year-end parting gift from the good folks at NIAID, Anthony Fauci's old fiefdom at the National Institutes of Health. NIAID – the same unaccountable and secretive agency that Fauci used to fund the gain-of-function research of Ralph Baric at UNC Chapel Hill and the Bat Lady in Wuhan that resulted in Covid – has a new director, one Dr. Jeanne Marrazzo.

Marrazzo and another NIAID colleague, Dr. Michael G. Ison, wrote a year-end editorial in the *New England Journal of Medicine* that accompanies a research paperon recent H5N1 Bird flu cases in the United States, as well as a case report of a lone case of severe illness associated with Bird flu in British Columbia.

Marrazzo and Ison summarize the findings of the research paper and case report as follows:

Investigators now report in the Journal a series of human cases from the United States and Canada. The former series involves 46 case patients with generally mild, self-limited infection with [Influenza type] A(H5N1): 20 with exposure to poultry, 25 with exposure to dairy cows, and 1 with undefined exposure....Most case patients presented with conjunctivitis, almost half with fever, and a minority with mild respiratory symptoms, and all recovered. The only hospitalization occurred in the case patient with undefined exposure, although hospitalization was not for respiratory illness.

They elaborate on the single case of serious illness:

In Canada, a 13-year-old girl with mild asthma and obesity presented with conjunctivitis and fever and had progression to respiratory failure...After treatment that included oseltamivir, amantadine, and baloxavir, she recovered.

In other words:

- Over an eight-month period, from March to October 2024, 46 cases of human bird flu occurred in the United States, a country of 336 million people.
- There were zero deaths.
- 45 out of 46 infected persons had known exposure to animals.
- The majority of the cases consisted of conjunctivitis (commonly known as "pink eye").
- Only one US patient was hospitalized, but this was not due

to pneumonia – the principal life-threatening complication of influenza – and the patient recovered.
- One severe case was identified in Canada, a country of 40 million people, in an asthmatic, morbidly obese girl. She was treated successfully with respiratory support and existing antiviral medications, and she recovered.

Does this sound to you like a public health emergency worthy of the legacy media's recent exhumation of discredited Covid-era fear-mongers like Dr. Leana Wen and Dr. Deborah "Scarf Lady" Birx? Does it justify their hair-on-fire pronouncements on cable news shows everywhere, pushing for indiscriminate PCR testing of animals and emergency authorization of more mRNA vaccines for humans?

Does this sound to you like justification to continue to kill and destroy (pro tip: "cull" means kill and destroy) millions upon millions of farm animals, when most animals who contract Bird flu survive, recover, and develop immunity?

Does this sound to you like justification for another Emergency Use Authorization of another mRNA vaccine?

No? Me neither.

But wait, there's more.

In their editorial, NIAID experts Marrazzo and Ison fail to mention the following:

- There have been zero cases of human-to-human transmission of this virus.
- The current circulating clade of the virus has been determined by independent researchers to very likely have originated at a US Government gain-of-function laboratory, namely the USDA Southeast Poultry Research

Laboratory (SEPRL) in Athens, GA.
- Multiple bioweapons laboratories, including the Yoshihiro Kawaoka lab at the University of Wisconsin, and the Ron Fouchier lab in the Netherlands (both of which have been affiliated with NIAID and with work done at SEPRL) have been doing gain-of-function research on Bird flu for many years, including experiments so outrageously dangerous that their work prompted President Obama's ultimately unsuccessful ban of gain-of-function research in 2014.
- In 2019, NIAID reapproved and resumed funding Kawaoka and Fouchier's dangerous work at increasing human transmissibility of Bird flu – the very same gain-of-function research that had prompted Obama's ban.
- According to its package insert, Audenz, the current Bird flu vaccine, was associated with death in 1 out of every 200 recipients, compared to 1 in 1,000 placebo recipients.
- According to openthebooks.com, and as reported in the *New York Post*, NIH scientists received royalties totaling $325 million from pharmaceutical companies and foreign entities over more than a decade.

So, what are our friends at NIAID's recommendations?

For one, they stress the "urgent need for vigilant surveillance of emerging mutations and assessment of the threat of human-to-human transmission."

Are they advocating for the willy-nilly testing of entire livestock herds, as promoted by Birx, which is sure to create a preponderance of false positives?

Are they calling for the continued mass killing and destruction of millions upon millions of farm animals, whenever a fraction of the animals test positive for the virus?

Instead of PCR-swabbing every cow, chicken, and farm worker on Earth, how about we stop creating new mutant variants of H5N1 in the labs, since that's where the current problem originated? How about we stop funding such utter madness with our tax dollars, funneled through corrupt government agencies like NIAID?

After all, you don't save Tokyo by creating Godzilla.

But Marrazzo and Ison make no mention of this common-sense, sane approach. Instead, they also stress the need for more – you guessed it – *vaccines*. They write:

> *we must continue to pursue development and testing of medical countermeasures…Studies have shown the safety and immunogenicity of A(H5N1) vaccines…studies are ongoing to develop messenger RNA–based A(H5N1) vaccines and other novel vaccines that can provide protection against a broad range of influenza viruses, including A(H5N1)."*

Aside from attesting to the "safety" of a product where 1 in 200 users die, the use of the word "countermeasures" is extremely telling. It is a military term, not a medical one. We have already seen this game played with Covid. The gain-of-function lab research is done to produce a lab-manipulated, weaponized version of a virus, a version that is transmissible among and toxic to humans – in other words, a bioweapon. The vaccine is the countermeasure to the bioweapon. The vaccine is the intellectual property of those who created the bioweapon, and it is worth a fortune once the weapon has been unleashed. It is as simple as that.

"Pandemic preparedness" is a gigantic, deadly protection racket. I have described it in the past as arsonists running the fire department. That is precisely what happened with Covid, and that is what is being attempted with H5N1 Bird flu.

Moving forward to a new administration that has expressed a commitment to rooting out corruption in the pharmaceutical/medical/public health realm, improving the health of citizens, and restoring trustworthiness in medicine, I recommend the following steps to combat the H5N1 Bird flu, and to end the "pandemic preparedness" racket that threatens to hold the world hostage again and again, as it did during Covid.

- Immediately end and outlaw all gain-of-function and other bioweapons research in and funded by the United States, and apply all possible diplomatic pressure to eradicate it from the Earth.
- Eliminate all special protections from liability for vaccines, including the 1986 National Childhood Vaccine Injury Act and the PREP Act.
- Refocus Infectious Disease research on new therapeutics, rather than power-seeking and profit-driven vaccine development.
- Completely reform the National Institutes of Health, and close the incorrigibly corrupt NIAID altogether.

The fear pornographers must be discredited. We must make realistic and sensible decisions about our food supply.

We must learn the lessons of Covid, and live in knowledge rather than in fear.

We must end the protection rackets, confidence games, and shakedowns that government insiders impose on us like mafiosi.

Happy New Year!

Caligula's Horse and the US Senate

Originally published on
November 3, 2022 by *American Thinker*.

According to the Roman historian Suetonius, the emperor Caligula made plans to install his favorite racehorse, a beast called Incitatus, as a Roman senator — although, as the story goes, Caligula was assassinated before he could do so. This and other deranged actions by Caligula have led some to hypothesize that he suffered some severe neuropsychological injury, in part because his behavior grew drastically unhinged after he narrowly survived a life-threatening illness.

In the case of John Fetterman, the stroke victim and current Democrat candidate for senator from Pennsylvania, the Democratic Party has done its utmost to match Caligula's arrogance, absurdity, and disdain for the citizenry, and to surpass the famously depraved emperor in cruelty.

Thanks to the combined efforts of the Democratic Party,

Fetterman's wife Gisele, one Dr. Clifford Chen, and the legacy media, the pathetic spectacle that was Fetterman's attempt to debate Dr. Mehmet Oz on October 25 was the first and only opportunity for most voters to see that John Fetterman is severely cognitively impaired, and beyond question unfit to represent them in the United States Senate.

Why would an emperor want to make a senator out of his horse? Why would a modern political party seek to do the same with a man who can neither speak coherently nor comprehend spoken language without electronic prompting?

Before any false outrage ensues, let me be clear: I am not suggesting that John Fetterman is an animal. Rather, he is a sick and severely impaired human being, and he should be getting proper treatment. He should not be used to grab political power while being publicly humiliated in the process. But this is exactly what the Democrats are doing to him now.

All humans, including John Fetterman, should be treated with respect for their fundamental human dignity. But human dignity is not the Democrats' strong suit these days — witness the skyrocketing crime, homelessness, and violence in *Democrat*-run cities and states; the unimpeded trafficking of children and fentanyl across the southern border under our current *Democrat* administration in Washington, D.C.; and the strident demands of *Democrats* nationwide for abortion on demand, without restriction.

Assuming that Caligula was not completely insane, one can surmise the intended message of his equine stunt: I can do whatever I want, however cynical, absurd, and degrading that may be. What matters is not the office, its intended function, or even that of government as a whole. All that matters is that I possess power and that no one questions my power.

What kind of political monsters would treat our government

and citizens as cynically as Caligula did his? Who would deceptively promote such a deeply impaired (and therefore easily manipulated) candidate as a competent individual for a key post in a representative democracy? What degree of imperious immorality would such behavior require?

With Fetterman, the horse is now out of the barn, so to speak, and the Democrats and their army of media lackeys are frantically shutting the doors. No doubt they will say, at some point, that this was nobody's fault. Stuff happens, don't you know. The guy had a stroke, and circumstances took on a life of their own.

Well, no. Fetterman's stroke occurred several days before the primary, and the eventual second-place finisher was a viable alternative — in fact, a sitting Congressman. But the Democratic Party stuck with Fetterman and chose to hide his impairments.

Sound familiar?

Oh, yeah — they did the same thing in the 2020 presidential election, with the obviously demented Joseph Marionette Biden. And how has that worked out for America? Recession, runaway inflation, the disastrous bug-out from Afghanistan, millions of illegal aliens pouring in through Mexico, proxy war in Ukraine, an impending energy crisis, and more. But at least Ron Klain, Susan Rice, or whoever else is pulling the strings in the Oval Office still possesses power.

A pattern has emerged, and only a party as depraved as Caligula himself would adopt such a strategy to seizing and holding power. To repeatedly deceive the electorate into electing mentally impaired persons — cognitive Trojan horses — to high office, all for them to be controlled by Deep State apparatchiks and party *éminences grises*, is beyond reproach. It reeks of a disdain for representative government so great that it makes an outright mockery of it.

We cannot know if Caligula's bizarre actions were born of malice or lunacy. But the power-hungry party attempting to install John Fetterman in the United States Senate cannot plead insanity. They already did the same thing with Joe Biden in the presidency. There is too much method here to be madness. This is evil.

What is Joe Biden's Life Expectancy?

> Originally published on July 18, 2022
> by *American Thinker*.

In this article, written 2 years and 4 months before the November 2024 elections, I predicted that Joe Biden's dementia would be sufficiently obvious before that election that it would preclude his candidacy for reelection. Biden ended his candidacy for reelection on July 21, 2024, two years after this article was published.

President Joe Biden is suffering from dementia, and it's worsening before our eyes.

You know it, I know it, everyone knows it.

His handlers know it. Why else would they issue him absurdly detailed cue cards telling him when to stand and sit?

His vice president knows it. Kamala Harris explicitly noted

her future running mate's cognitive problems during the 2020 Democratic debates.

Even the president's drug-addicted, prostitute-frequenting, influence-peddling son knows it. Hunter cruelly mocks the elder Biden's senility on the Laptop from Hell.

It is a national tragedy that a fragile, senile, *non compos mentis* old man is the president of the United States. Nevertheless, this scandalous state of affairs is downplayed so much by the Democrats, the Washington bureaucracy, and complicit press and social media outlets that Biden and many around him — Harris included — still speak seriously about Biden seeking a second term in 2024.

Only very recently are cracks starting to form in the leftist elites' see-no-evil attitude to Biden's dementia. See the *New York Times'* recent article for evidence. However, it appears that this new view is predicated more by Biden's currently horrendous polling, which threatens to sink the Democrats' midterm prospects, than by Biden's cognitive decline, which has been evident for years.

The notion of a second Biden term raises another question about the man, one even more existential than his undeniable cognitive dysfunction. The question is, would Joe Biden even live through a second term as president?

As a practicing physician who cares for numerous elderly patients and addresses dementia and end-of-life issues on a regular basis, your humble correspondent decided to take a closer look at this question.

Here are my findings:

Joe Biden's listed birth date is November 20, 1942. Thus, at this writing, Joe Biden is 79 years and 8 months old.

Were he to win re-election in 2024, then by his second

inauguration, he would be 82 years and 2 months of age. At the end of a second term, he would be 86 years and 2 months old.

How long does the average 79-year-old man have to live?

According to the Social Security Administration, the life expectancy for a 79-year-old American male is an additional 8.82 years, for a total life expectancy of 87.82 years. This would take Joe Biden to late 2029 — less than a year after the end of a second term. That's cutting it awfully close, especially for a man struggling as badly as he obviously is right now.

But it gets worse.

Those Social Security numbers are for *all comers* in Biden's age group. But remember, Joe Biden has dementia, and dementia shortens life expectancy.

How much?

A recent study by Carol Brayne et al., published in the prestigious British Medical Journal, concluded that the median survival time from a diagnosis of dementia is 4.5 years for all persons and 4.1 years for men.

Assuming we very generously "diagnose" Joe Biden's dementia as starting now — say, July 1, 2022 — then 4.1 years places us around August 2026. In other words, by this measure, Biden's most likely date of death will be less than halfway through his second term as president.

I also reviewed the standard clinical scale for measuring progress of dementia, known as the Reisberg Scale or the Global Deterioration Scale (GDS).

The GDS divides dementia into 7 stages, Stage 7 being the most severe. By a careful review of the criteria and of available video of Biden over time, I believe he is currently at least at Stage 4, which is referred to as "moderate cognitive decline."

It is worth noting here that the President of the United States is one of the most supported and protected persons on Earth. Joe Biden makes limited public appearances, which are carefully scheduled, tightly controlled, and highly choreographed. We, the public, only see Joe Biden at his *best*. It is possible that his dementia has actually reached Stage 5 on the GDS, or "moderately severe cognitive decline."

Given his demonstrated ability to still make prepared statements and public appearances, it seems unlikely to me that Biden's dementia has advanced beyond Stage 5, where many basic mental functions are obviously impaired.

So what does this mean moving forward?

The cited average duration of Stage 4 dementia is two years. For Stage 5, it is 1.5 years. Again, if we very generously assume that Joe Biden is "entering" Stage 4 right *now* — say July 1, 2022 — then he is likely to progress through Stages 4 and 5 and into Stage 6, "severe cognitive decline," by late 2025, approximately a year into a second term.

In other words, a second Biden term presents the very real prospect of a president who cannot identify his family members, count down from ten, nor control toilet function, with three years remaining in his term.

An important side note: Joe Biden also has a serious problem with falls. Recurrent falls are a strong predictor of death in the elderly.

Biden's recent bike accident on vacation at Rehoboth Beach, his repeated stumbles climbing the stairs to Air Force One in March 2021, and his broken foot sustained while playing with his dog in November 2020 provide ample evidence that the president is a "fall risk." And of course, these are just the incidents known to the public. There may be many more.

Biden could have very easily broken a hip on any of the above occasions. When an elderly patient falls and breaks a hip, his mortality *within one year* is over 20 percent, even with surgery. Once again, dementia increases the risk.

Furthermore, Biden reportedly suffers from atrial fibrillation, an irregular heartbeat that puts the patient at high risk for stroke. For this he reportedly takes Eliquis, a blood-thinner that prevents stroke but significantly increases the risk of hemorrhage in the case of falls.

Patients requiring anticoagulation for stroke prevention who are also fall risks are caught between a rock and a hard place. A fall with head trauma while on "blood thinners" may cause severe bleeding in the brain. However, discontinuing the anticoagulation medicine causes their risk of embolic (blood clot–related) stroke to increase dramatically. Joe Biden appears to be in this situation.

In summary, after a detailed review of relevant medical data, here are this physician's conclusions:

On Election Day, November 2024, President Joe Biden will, more likely than not, still be alive. However, his dementia will most likely be sufficiently advanced by then that it will be obvious to everyone that he is incapable of serving another term.

By Inauguration Day, January 2029 (at what would be the end of a second term), President Joe Biden will, more likely than not, have died of natural causes.

Given the utmost importance of the Presidency, I implore the people of the United States to take these concerns more seriously than they have done until now.

Disaster: CDC Calls for Covid-19 Vaccines on the Child Immunization Schedule

Originally published on October 28, 2022 by American Thinker.

The CDC Advisory Committee on Immunization Practices (ACIP) met on October 20, 2022. Despite the protestations of child advocates and physicians, including your humble correspondent (see Appendix A), the committee voted *unanimously* to add the Covid-19 vaccines to the Child and Adolescent Immunization Schedule.

Put bluntly, the committee blew it. Scientifically, ethically, and as public policy, this was a terrible decision — and one that, if allowed to stand, will cause irreparable harm to countless American children.

Here are the top ten reasons that Covid-19 vaccines should *not* be on the Child and Adolescent Immunization Schedule:

1. **The risk/benefit ratio of the Covid-19 vaccines is UNFAVORABLE for children.** It is now known beyond debate that the risk of serious illness and death from Covid-19 a) is strongly correlated with advancing age, b) is declining over time as the virus continues to mutate, and c) is and always has been vanishingly small in children. Comparison to VAERS and EudraVigilance data shows much greater risk to children from Covid-19 vaccine-adverse events than from Covid-19 infection itself.

2. **The Covid-19 vaccines do NOT stop a recipient from either contracting the disease or transmitting it to others.** Exhibit A for this would be the recent Covid-19 diagnosis of the reportedly quintuple-vaxxed CDC director, Rochelle Walensky. These vaccines are ineffective in efforts to create herd immunity — unlike many of the established vaccines for other diseases that are already on the schedule.

3. **The CDC decision ignores natural immunity.** The CDC *itself* has released recent data showing that 86% of school-aged American children already have had prior Covid-19 infection. With the committee's decision, the CDC *yet again* dismisses natural immunity as a protective factor against Covid-19 disease — a position so unscientific and intellectually dishonest that other factors (political, monetary) seem the only plausible explanation for it.

4. **The Covid-19 vaccines have NO long-term safety data.** How could they? They have been in use less than

two years. Young children receiving these vaccines will live for up to a century having to deal with the unforeseen risks of these injections, the magnitude of which we have zero knowledge of at present.

5. **The Covid-19 vaccines have much more worrisome safety data than vaccines already on the schedule.** Covid-19 vaccines have well known, well demonstrated, significant toxicities in children and young persons (e.g., myocarditis) that occur in significant excess relative to other vaccines currently on the Child and Adolescent Immunization Schedule. Reports of vaccine-related deaths from Covid-19 vaccines far outnumber reports related to other time-tested vaccines already on the CDC schedule.

6. **The CDC's decision completely contradicts those of multiple nations who have made appropriate risk/benefit analyses and ended pediatric Covid-19 vaccine programs.** Sweden, the rare nation that remained mostly sane in its Covid-19 policies since the onset of the pandemic, never recommended Covid-19 vaccines for children under 12 and has now ceased to recommend them in older children as well. Denmark no longer offers Covid-19 vaccines to any healthy children under 18. The CDC, to my knowledge, has never publicly addressed the decisions of these nations or why they were made.

7. **Adding Covid-19 vaccines to the CDC schedule will absolutely result in their being required for school attendance in many states.** The CDC has, rather disingenuously, stated that their schedules are not

mandates. However, the Covid-19 era has taught us that most health commissioners and school officials, especially in Democrat-run states, treat CDC recommendations as gospel. Gavin Newsom in California has already stated he will mandate Covid-19 vaccines for schools in California. Kathy Hochul in New York and Gretchen Whitmer in Michigan will no doubt do the same if re-elected. These recommendations are effectively mandates for millions of American children — approximately 13 million in California and New York alone.

8. **The decision is completely out of step with most American parents.** The CDC's own data show that most parents have chosen *not* to have their children vaccinated against Covid-19. Furthermore, the younger the age group, the lower the percentage of children vaccinated. And somehow, these children have survived Covid-19! In essence, parents have collectively shown a degree of logic, common sense, and ability to perform basic risk-benefit analyses that seems completely beyond the capacity of the CDC. The only remaining question for many thinking, concerned parents seems to be whether the CDC a) doesn't give a damn about children, b) is completely incompetent, c) is completely corrupt, or d) all of the above.

9. **Mandating these new and minimally tested therapies on millions of American schoolchildren will further undermine trust in the immunization schedule for years to come.** In short, it will backfire. Population-wide confidence in the current vaccine schedules has already

been seriously undermined by the Covid-19 vaccine policies that have caused countless Americans (including thousands of health care workers) to lose their jobs. In my practice, I now find a number of patients refusing some or even all vaccines, despite my recommendations otherwise, as a result of the coercion they felt regarding Covid-19. This will happen at the pediatric level, as well, if Covid-19 vaccines are imposed on schoolchildren. The horrible irony is that children will suffer and die for lack of beneficial vaccines as a result of the CDC's attempts to force unnecessary Covid-19 vaccines on them. This is public health malfeasance of the worst kind.

10. **The prime directive of medicine is** *primum non nocere* — first, do no harm. At present, the risk/benefit ratio for the Covid-19 vaccines in children is not even close. The CDC committee's decision gives the middle finger to all four pillars of medical ethics: autonomy, non-maleficence, beneficence, and justice. For this reason alone, it should have never been made.

The CDC Advisory Committee on Immunization Practices' recent unanimous decision to add Covid-19 vaccination to the Child and Adolescent Immunization Schedule is shameful. It is unscientific, unethical, bad policy. It is further evidence — if we need any more at this stage in the Covid-19 saga — that the public health bureaucracy in the United States is corrupt and broken, perhaps beyond repair.

The potential electoral defeat of gubernatorial Covid-19 tyrants like Kathy Hochul or Gretchen Whitmer would provide respite to millions of families and children, and would be most welcome to the vulnerable in those states. However, the dismantling of the medical-industrial complex in Washington, D.C. is the only ultimate solution to the seemingly endless governmental assault on medical autonomy that the Covid-19 pandemic has made so obvious.

A Voter's Primer: The Seven Health Policy Habits of Insanely Progressive People

Originally published November 7, 2022
by *American Thinker*.

What is the most painful lesson the Covid-19 saga has taught the American people? Perhaps it was how, over the past two and a half years, we have been force-fed an extended taste of the dystopian version of "health care" that the Democratic Party and the Washington nomenklatura (but I repeat myself) seek to permanently impose on us.

We've lived through it now, so we can't say we weren't warned. However, many people are still dazed and confused, others remain terrified, and some just want to forget that the whole nightmare ever happened and return to their old pre-pandemic lives.

Things may have calmed down, but don't kid yourself: the good old days have not returned. It's vital that we never forget what they did to our children, our livelihoods, and our civil rights, how they ruined countless lives in the name of "keeping us safe." And they're not done. They're just warming up. And now they seek amnesty for their abuses of power, even as they continue to persecute courageous dissidents like Dr. Peter McCollough, even as they push booster after booster, even to tiny children? No way.

In the interest of jostling the collective memory before this election, your humble correspondent asks: what are the medical implications of Democratic Party governance?

Presenting the Seven Health Policy Habits of Insanely Progressive People:

1. Martial law as "public health." Remember how "Two weeks to flatten the curve" became two years to flatten your will to live? Get ready for draconian, indefinitely extended lockdown measures whenever a "public health emergency" is declared — for example, before major elections.

2. Mandatory jab policies gone wild. This isn't just your father's measles-mumps-rubella vaccine, folks. We're talking about warp-speed-produced, novel-technology shots with *zero* long-term data. You think they're stopping with SARS CoV-2? Have you noticed how much they're talking up RSV? There's a *lot* of money to be made. You think you'll have a choice, right? After all, they're all about "My body, my choice," right? Well, think back to about a year ago. Did the shots do what they claimed they would? Did they apologize after vilifying and persecuting the skeptics? If the Dems keep power, the questions return: do you want to earn a living? You will comply. Do you want freedom of movement? Comply. Do you want your kids to go to school? Comply.

3. Censor and destroy all dissenting physicians, scientists, and health care workers. Remember NIH chief Francis "Over the

Rainbow" Collins's call for a "quick and devastating published takedown" of the so-called "fringe" scientists (From Harvard, Stanford, and Oxford!) who wrote the Great Barrington Declaration? Fast-forward to the fascistic current attempts to strip Dr. Peter McCollough of his board certification. Welcome to a world of compromised, careerist medical mediocrities beating down pre-eminent minds who refuse to keep silent, like Orwell's boot stomping on a face forever and ever.

4. Support and grow the Government-Pharmaceutical-Industrial Complex. One really should read *The Real Anthony Fauci* by Robert F. Kennedy, Jr. to get the full picture. The huge problem of regulatory capture, the intertwining of U.S. government medical agencies with the military (especially with regard to vaccine development), and the absolutely massive amounts of money involved are truly head-spinning. Only the terminally naïve or willfully blind could believe that Fauci et al. care one whit about the individual citizen's well-being.

5. Predatory medical and social policies against children. Shuttering schools for two years at a time. Masking young children and toddlers. Still — to this day — pressing for mandated Covid-19 vaccines for schoolchildren. Population-wide psychological trauma as well as developmental and educational delay, all resulting from their cruel, excessive, and utterly unnecessary policies. Any apologies? New York Democrat governor Kathy Hochul still "reserves the right" to bring back masks. Still not convinced? Need I mention Drag Queen Story Hour? Or abortion on demand as a secular sacrament? Why do Democrats hate children so?

6. The woke Lysenkoism of academic medicine. The November 1, 2022 issue of the once-pre-eminent *Journal of the American Medical Association (JAMA)*, published one week before the midterm elections, was almost completely devoted to attacking the *Dobbs* Supreme Court decision. Curiously, it contained only two original research articles — neither one about abortion. However, interspersed with the full-page Big Pharma advertisements, the issue had no fewer than nine opinion pieces, all pro-abortion and anti-*Dobbs*. No contrasting views permitted. Journals that used to print original research and promote debate of controversial issues now print propaganda better suited to a Planned Parenthood brochure. And medical schools' curricula are no better.

7. Complete government control of all aspects of medicine. Forget about the old bugbear of "socialized medicine." With Medicare, Medicaid, Obamacare, federal medicine, and NIH influence over academic medical centers, American medicine is already socialized. However, during Covid-19, we saw government manufacture and enforce complete consent to its health care policy at a level never seen before. This was accomplished by thorough capture of hospital systems and local health officials with a crude but effective carrot-and-stick approach. Do exactly what we say, and we pay you off handsomely down the road (with taxpayer dollars). Don't do what we say, and we shut you down. For good.

To those uncertain about the relative merits of communism versus capitalism, I often say: "Be honest with yourself. Which Korea would you rather live in: North or South?" To anyone unsure how to vote in this upcoming election, I now ask, "Be honest with yourself. What type of America would you rather live in for the rest of your life — one like Covid-era Florida or one like Covid-era New York?"

Neither American political party is perfect — far from it. But one party never wants things to go back the way they were before Covid-19. Never.

Vote wisely, America.

Covid-19 is Becoming Milder, but the Left Stays Toxic as Ever

Originally published on April 22, 2022
in *American Thinker*

In this article, I contrasted how a new virus mutates, evolves, and adapts to coexist with its host – a remarkably reasonable negotiation for a parasite, one must admit. In so doing, viruses compare favorably with the mainstream media and the far left (but I repeat myself), who never seem to relent or relax their malevolence in the pursuit of peaceful coexistence.

Just this morning (Tuesday, April 22, 2022), in the *New York Times*' "the Morning" online report, a guy named David Leonhardt writes with apparent amazement that "Coronavirus cases have risen in major cities. Hospitalizations have not." Imagine that.

Leonhardt goes on to note that despite the long list of members of Congress and other public servants recently diagnosed with

Covid-19, none of them, even our superannuated speaker of the House, appears to have got very sick from it. To his credit, he supplements this observation with some charts that clearly show the disconnect between current cases (which are rising) and hospitalizations (which remain flat).

So far, so good. But then he gives his explanation for this trend. That's where the spin and outright dishonesty of the Covid-forever left — led by the *Times* — continue apace.

To what does David the Lionhearted, the Gray Lady's intrepid knight of the keyboard du jour, attribute these positive trends? He reports what (supposedly) "many experts believe:"

- Vaccines and booster shots are effective and universally available to Americans who are at least 12. (Covid [sic] continues to be overwhelmingly mild among children).
- Treatments — like Evusheld for the immunocompromised and Paxlovid for vulnerable people who get infected — are increasingly available.
- Tens of millions of Americans have already been infected with the virus, providing them with at least some immunity.

Two key points should be drawn from this list of explanations.

First, an absolutely central reason for the good news about Covid-19 has been deliberately omitted. Second, several systematic lies are embedded in the three explanations that are given.

As any truly knowledgeable and honest doctor or virologist — provided you can find one these days — will tell you, viruses such as SARS-CoV-2 mutate like crazy and evolve rapidly and *in a predictable manner*. In short, these viruses consistently mutate to become more transmissible and less virulent. Why do they evolve in this way? For the same reason all organisms evolve: to benefit their own propagation and survival.

When a new virus is first introduced to a host species, the initial interaction is often not pretty. The virus may struggle to spread between individuals, endangering its survival, and it may invoke severe illness in its host, even killing it, thereby endangering *both* species' survival.

Moving slowly and painstakingly from one home to another, while burning down the one in which you currently reside, is no way to survive. So the virus mutates and evolves into a milder form that spreads more readily yet sickens the host less.

In essence, the perfect respiratory virus is the common cold. It infects its host but makes the host sick enough only to sneeze the virus's progeny at everyone around. It spreads like wildfire, but it doesn't burn down its own house in the process.

Not for nothing, but what do the other coronaviruses in general circulation among humans cause? That's right: symptoms of the *common cold*. This is almost certainly the final common pathway for SARS-CoV-2.

As a practicing physician, trained before schools of public health veered to the left of gender studies departments, I have been saying this since the summer of 2020.

Meanwhile, panic pornographers ranging from Anthony Fauci to *Times* newsboy Leonhardt's "experts" have latched onto that first trait of viral evolution (increased transmissibility) while deliberately downplaying, or even denying the second (reduced virulence). Why? Because they want to foment all the fear that increased transmissibility promotes, yet allow none of the hope and perspective about the virus that acknowledging reduced virulence would bring.

The second lesson to take from Leonhardt's list is this: as the facts become too obvious to support their false narrative, leftists perform the propagandistic equivalent of a "tactical retreat," covering their tracks with false and misleading explanations.

Leonhardt writes that "vaccines are effective and readily available." Effective at what? At stopping the virus in its tracks, as the *Times* and Fauci claimed for months? Nope. At preventing persons from contracting Covid-19, as they also claimed? Well, obviously not, since every one of those politicians has been vaccinated and boosted to the hilt. At reducing severity of disease? Well, then what happened to the vaunted "pandemic of the unvaccinated?" Based on the data curves Leonhardt provides, the unvaccinated aren't going to the ICU these days, either.

Leonhardt touts Evusheld and Paxlovid as "increasingly available," a total *non sequitur* in the absence of any data supporting their role in the current trends, which he does not provide. He completely ignores any cheap, repurposed early treatments, despite — or more likely because of — the growing mountains of data supporting their effectiveness.

Finally, Leonhardt blatantly understates the effect of natural immunity, both by lowballing the number of previously infected Americans (it's in the *hundreds* of millions, Dave, not tens) and by the deeply misleading statement that prior infection produces "at least some immunity" (Natural immunity is far superior to vaccine-related immunity.)

Here is the reality, the plain fact that Fauci and Leonhardt's "many experts" will never admit: SARS-CoV-2 is evolving and adapting to coexist with us. The Covid-forever left remains as toxic and destructive as ever.

The Top 10 Covid Villains of 2021

Originally published on January 22, 2022
by *American Thinker*.

Written shortly after the one and only known case of Covid I've had, contracted at the height of omicron, I wrote this "Top 10 List" as a humorous catalog of the worst Covid villains, at least as I saw them at the time. The list still seems pretty solid to me, although some regional bias is present.

So much more has come to light since, and some of these people are probably minor players in retrospect. Still, the wicked tend to do evil on the scale that the circumstances allow, and it's safe to say no one on this list was falsely accused.

Your humble correspondent hoped to have this retrospective completed by the start of the New Year, but a tussle with Omicron intervened. Nevertheless, it still seems worthwhile to pause and remember the worst Covid offenders of 2021.

As we do so, questions inevitably arise. When will these wrongdoers be held accountable for the death and destruction they have wrought on our world? What should their punishments be? How should they be made examples of — cautionary tales to dissuade those who might plot similar evils in the future?

Without further ado, here are our Top 10 Covid-19 Villains of 2021.

10. Sonny and Fredo Cuomo (tie). Of the two fatally flawed Cuomo boys, Andrew is the hotheaded, bullying, enemy-collecting former Don, while Chris is the hapless beta who got dropped on his head or something. Before Andrew paid the ultimate price for his sexual harassment escapades, he managed to whack thousands of innocent yet expendable nursing home nonnas. Chris ran interference for his big brother on CNN, then briefly acted as his damage-control consigliere until he got the kiss of death himself, leaving the Cuomo family in desperate need of a Michael.

9. Randi Weingarten. The Jimmy Hoffa of "educators," Weingarten is president of the American Federation of Teachers, the most powerful teachers' union in the United States. More than any other single individual, Weingarten spearheaded the prolonged and utterly unnecessary school closures that shattered the lives of millions of children throughout the Covid-19 era. Now she attempts to gaslight the public about her central role in that titanic injustice, while still pushing for school closures during Omicron. That's the kind of feral, black-hearted goblin that makes Grendel's mother look like Donna Reed.

8. Kathy Hochul. When New York's current governor replaced the execrable Andrew Cuomo, citizens thought better days must lie ahead. Instead, in a classic "hold my beer" moment, this unknown quantity simply changed the motto above the governor's office from "GRABBIN' FANNIES AND KILLIN' GRANNIES" to "VAX OR DIE." With her creepy-as-all-get-out *Vaxed* necklace, her remarkable resemblance to Margaret Hamilton, and her demonic lust to oppress the Oompa Loompas formerly known as New Yorkers, Hochul has forever claimed the title of the "Wicked Witch of Western New York." Giving this devil her due, Hochul is a bit of a phenomenon: a truly dangerous harpy who apparently sprang fully formed out of the primordial ooze that is local New York Democratic Party politics.

7. **Rochelle Walensky.** To some, Dr. Hot Mess may seem to have a bit too much Lucy and Ethel in her to make this exclusive list. Hardly the sharpest knife in the drawer, Walensky (Biden's choice for CDC head) *is* a bit of a dark horse, but hear us out. Walensky so brazenly seeks to abuse power that she attempted to singlehandedly prolong the federal eviction moratorium on public health grounds — giving Americans a shot across the bow about how woke Big Medicine feels about *all* our civil rights. She has also driven the CDC's remaining credibility off a cliff with 1) the laughably bad in-house studies she has used to back her recommendations, 2) the email revelations that she takes policy orders from Weingarten's teachers' unions, and 3) her utterly incoherent change-with-the-wind "guidance."
Machiavelli warned against assuming malice when incompetence explains abuse of power, but Walensky proves it's a false choice when the person in question is so generously blessed with both. Bonus points for coming so late to the game, knowing exactly the evil she was signing up for, and diving in headfirst.

6. **Francis Collins.** The human embodiment of hypocrisy, Collins presents himself as Mister Rogers while behaving like Dr. Evil. The ultimate enabler, front, and (relatively) silent partner to Anthony Fauci for the past dozen years, Collins retired as NIH director after being caught lying to Congress about the gain-of-function research his NIH funded at the Wuhan Institute of Virology, which, as everyone from Rand Paul to Jon Stewart knows, was the source of this whole disaster.

Recently FOIAed emails reveal that Collins, notwithstanding his fatuously gentle, avuncular public persona, demanded a "quick and devastating published takedown" of the three "fringe epidemiologists" (from Harvard, Oxford, and Stanford!) who wrote the Great Barrington Declaration. How *dare* such insolent (if truly expert) commoners attempt to avert the immense, worldwide collateral damage that Lords Collins, Fauci, and Co. have decreed to befall us!

When the final analysis is made about the origins of Covid-19 and the harms deliberately caused by the official response to it, Collins's sins will closely shadow Fauci's. In one respect, though, Fake Francis stands alone: without question, he is the most nauseatingly disingenuous personality on this list. Just watch his self-serving bastardization of "Somewhere Over the Rainbow" to see this human gag reflex's repulsive personality in full flower.

5. **Albert Bourla.** Ever wonder why the push for mass vaccination with mRNA vaccines makes you feel like one head in a herd of cattle during a brucellosis outbreak? Well, wonder no more. That's right, ladies and gentlemen: the CEO of the world's largest and most rapacious pharmaceutical company is a *veterinarian*. But there's no *All Creatures Great and Small* vibe coming from this guy; it's all human beings as livestock, all the time.

Bourla reportedly made $21 million in salary alone from Pfizer in 2020, while publicly calling all those who questioned the universal, forced, and repeated administration of his leaky, toxic, rapidly obsolescing vaccines "murderers." Project much? The amoral poster child for Big Pharma and its evil spawn, Big Vax.

4. **Mark Zuckerberg.** He's the smarmy, impudent personification of social media censorship, population-level thought control, and exploitation of the masses. Since Day One, Facebook has been the epicenter of suppression and silencing of all dissent from the lockdown/no early treatment/vaccines forever Covid-19 narrative.

Zuckerberg's foundation — in concert with J&J's — even funded the putrid *Atlantic* hit piece on Dr. Robert Malone. You can almost see Zuck rubbing his hands together with perverse pride, knowing he has finally made the jump from mere arrogant social media über-jerk (à la Jack Dorsey) to a genuine Silicon Valley force for worldwide evil, like...

3. **Bill Gates.** The *éminence grise* behind the whole disaster, and damn proud of it. Gates has wielded the obscene wealth of his Foundation as a cudgel, hijacking effective control of the World Health Organization's policy and messaging.

Furthermore, he partnered with Anthony Fauci to promote and produce the perverse fear-mongering infectious disease war games such as MARS 2017, SPARS 2017, and Clade X. His goal: To condition government agencies, the legacy media, and the medical establishment to the inevitability of just this sort of pandemic, and the need for the totalitarian measures and forced vaccination programs that have since come to pass. Pure evil.

2. Xi Jinping. As the pandemic began in his own country, Dear Leader responded by 1) brutally locking down his own people, 2) systematically lying to the rest of the world, and 3) ensuring the spread of the virus worldwide. Two years later, he's locking his people down again to ensure that his "genocide Olympics" won't be impacted.

Xi might have been ranked first but for two setbacks. First, he looks too much like Winnie the Pooh. Second, Xi *is* a communist dictator, after all. He's *supposed* to be evil.

1. Anthony Fauci. The linchpin of the entire Covid-19 disaster, Lil' Tony makes Dr. Strangelove seem like Marcus Welby, M.D. Fauci has been a central antagonist at every step of this global catastrophe, from funding its origins in Wuhan to the allowance of its worldwide spread to the systematic suppression of existing treatments to all the repressive policies to the takeover of the medical profession by Big Vax to the effective revocation of the Bill of Rights by the Deep State.

But don't take our word for it. Read RFK Jr.'s *The Real Anthony Fauci*, and ask yourself this question: what fate does this individual deserve if just one-tenth of that book is true? In our humble opinion, Fauci will be remembered as the greatest criminal against humanity since Pol Pot, and the most notorious traitor in the U.S. federal government since Alger Hiss.

Honorable mention: J. Biden, J. Dorsey, B. Johnson, R. Klain, S. Morrison, G. Newsom, the Scarf Lady, and the Bat Lady.

There you have it, folks: our top 10 Covid-19 villains of 2021! Did we forget your favorite? Please let us know in the comments below.

Happy 2022!

Covid-19 in 10 Sentences

Originally published on December 31, 2021
in *American Thinker*.

Written as a year-end, farewell article for the year 2021, this piece published in American Thinker gathered some acclaim as a succinct summary of the Covid era. It also marked my first use of the "ten sentences" structure I've since adopted as an occasional stylistic technique.

Reading it now, it is striking to see how much one could surmise even then about the wicked machinations underpinning Covid. And yet, so many people still don't realize today what was done to them and why.

As we approach the end of *annus horribilis* 2 (also known as 2021 A.D.), it seems worthwhile to to look back and summarize the events that have brought us where we are in the Covid-19 saga.

Here, in ten sentences, is how we got here.

1. Since at least 2014, the U.S. National Institutes of Health (NIH), through Anthony Fauci's NIAID division, have sent millions of U.S. tax dollars to communist China to fund research involving the genetic alteration of coronaviruses at the Wuhan Institute of Virology.
2. Around October 2019, the Covid-19 pandemic began when a new coronavirus leaked out of the same Wuhan Institute of Virology and into the human population.
3. The Communist Chinese Party imposed a tight lockdown of its own population, while simultaneously allowing international travel to and from China, facilitating the virus's worldwide spread.
4. As the pandemic unfolded, public health officials and the media used grossly overestimated death rates and false promises of self-limited measures ("Two weeks to flatten the curve") to promote unprecedented policies of prolonged, widespread quarantine of healthy populations, which continue to this day — two years later.
5. Simultaneously, in places such as New York State under former governor Andrew Cuomo, authorities knowingly put sick Covid-19 patients into close contact with highly vulnerable persons such as nursing home residents, resulting in tens of thousands of unnecessary and avoidable deaths.
6. Despite definitive evidence from the early stages of the pandemic that Covid-19 poses minimal risk of severe illness and statistically zero chance of death in children, and that children are not significant drivers of its spread, the Democratic Party and the public teachers' unions — with the help of health officials and the mainstream media — have forced schools to close for in-school learning for multiple school years, and continue to push for renewed

school closures in many areas of the country.

7. As cheap, existing, and safe medications and treatments were identified that showed effectiveness in treating Covid-19, a systematic, worldwide movement to suppress and discredit such treatments was instigated by Anthony Fauci, Bill Gates, the mainstream media, Big Pharma, and social media corporations, to protect their financial interests in vaccines and other proprietary medicines they had in development, resulting in tens of thousands of unnecessary deaths.

8. As Covid-19 vaccines became available in the U.S. through Emergency Use Authorization (EUA) from the FDA, these extremely new treatments were heavily promoted by Fauci, Gates, the media, Big Pharma, and social media under knowingly false pretenses, including repeated false claims that the vaccines 1) would provide herd immunity, 2) were equal or even superior to natural immunity, 3) stopped contraction and transmission of the virus, and 4) were safe and effective for all ages.

9. Even as the Covid-19 vaccines have now been shown to 1) lose effectiveness in a matter of weeks; 2) be ineffective at stopping transmission and spread of the virus; and 3) be inferior to natural immunity, and even as more than 20,000 vaccine-related deaths have been reported in the CDC's own Vaccine Emergency Reporting System (VAERS) — with a similar level of reports in EudraVigilance (the E.U.'s reporting system), the likes of Fauci, President Joe Biden, current New York governor Kathy Hochul, and New York City mayor Bill de Blasio continue to press ever harder for repeated doses of these same vaccines, including among young children.

10. Although the current dominant strain of Covid-19 — the Omicron variant — has been demonstrated to be more

transmissible and much less deadly than prior strains, as well as dramatically mutated from the original strain after which the vaccines were modeled, Fauci, the Biden administration, the Democratic Party, and the mainstream media are now employing a policy of endless boosters with the increasingly obsolete yet lucrative vaccines, alongside the systematic scapegoating of unvaccinated persons, rather than employing the focused protection of the vulnerable and promotion of normal life and natural immunity among the healthy that has already been successfully implemented in numerous "free" states.

What conclusions can we draw from this series of events? Here are a few:

First, the "health care industry" is largely a syndicate run by government bureaucrats like Tony Fauci and Francis Collins, Big Pharma, and ultra-rich investor-influencers like Bill Gates.

Second, the mainstream media and major social media platforms like Google, Facebook, and Twitter are diametrically opposed to freedom of speech and the free exchange of ideas. In fact, their goal is the opposite: an Orwellian thought control of the population and the suppression of all dissenting voices.

Third, the Democratic party is utterly corrupt and power-hungry, while the Republican Party is hopelessly gutless and ineffective.

Lastly, the formula has been revealed for the permanent extinguishing of the civil liberties outlined in the Bill of Rights: declare an emergency, terrify the populace, control the message, stifle all dissent, and revoke the citizens' freedoms indefinitely, all while grabbing and consolidating political power. Coming soon: the climate "emergency."

Happy 2022!

No, seriously: Outgoing NIH director picks up guitar and sings 'Covid over the Rainbow'

Originally published December 18, 2021
in *American Thinker*

Francis Collins, the outgoing Chief Pharisee of the National Institutes of Health, gave a bizarre farewell address on December 14, 2021, during a so-called U.S. Department of Health and Human Services "town hall" event (which, despite the name, appears to have been pre-recorded on a soundstage).

In so doing, Collins achieved a rare feat, something generally unachievable by mortals. Like one of the ancient Greek gods, he managed to completely and perfectly personify a particular aspect of human behavior. As Aphrodite personified sexual desire and Athena wisdom, so in this special moment did Francis Collins become the quintessential embodiment of...hypocrisy.

Francis Collins is the man who oversaw Tony Fauci and the NIAID's funding of coronavirus gain-of-function experimentation (yes, it *was* gain-of-function research) at the Wuhan Institute of

Virology (yes, it *was* the place where Covid-19 originated). As Rutgers professor Richard Ebright put it:

> *The NIH — specifically, Collins, Fauci, and [Lawrence A.] Tabak — lied to Congress, lied to the press, and lied to the public. Knowingly. Willfully. Brazenly.*

Collins made his career by supporting and funding morally objectionable research of many kinds while sporting the public persona of a sort of hyper-spiritual and unctuously tenderhearted Christian. Nowhere was this on fuller display than at this "town hall."

In a performance that exemplifies the term "tone-deaf" in so many ways, the dear, avuncular Collins picked up a guitar — complete with a fake expression of surprise at its presence on the set — and gave a birthday party clown's rendition of his own clumsy, cringe-worthy parody of "Somewhere Over the Rainbow," entitled "Somewhere Past the Pandemic."

You have to see it to believe it.

With his gray walrus mustache, Francis resembled an underfed Captain Kangaroo. His fingerpicking was rudimentary, his rhythm variable, and his voice alternated between a thin warble and abrupt talk-singing. He channeled Fred Rogers as best he could, albeit with none of the sincerity and ten times the creepiness. Through it all, he carried a smarmy glint in his eye that seemed to say, "See? I didn't have to be an evil scientist. I could've been a third-rate folk singer!"

But it was the words, ladies and gentlemen, the *words*. They weren't lyrics; there was nothing lyrical about them. Yes, they were badly enough composed — extra syllables at the end of several lines, strained rhymes like "come off" with "cough" — to send Harold Arlen and Judy Garland spinning in their graves.

But it was their message that sent Francis into his Apotheosis of Hypocrisy. For example:

"Somewhere past the pandemic, we'll hug our friends / And thank the people and science that brought the pandemic's end."

In the seemingly benign guise of an amateurish, lighthearted children's singer, Francis Collins achieved several Olympian feats of hypocrisy.

He made a cartoon of the misery and death suffered by millions of people worldwide, caused by a virus whose origin he almost certainly played a role in, and also caused by the draconian measures; suppressed therapies; and rushed, mandated vaccines that he, Fauci, and the rest of his NIH foisted upon the world.

He attempted one last time to gaslight the world about his role in this great disaster, and he recast himself and his cronies as grandfatherly benefactors of mankind rather than the amoral profiteers that they are.

He played his utterly phony public persona to its logical extreme, one last nauseating time.

The whole song lasted about two minutes, though it seemed much longer. Finally, *finally*, struggling to reach the top notes, Francis ended with:

"Let's...end...Covid...now!"

Yes, let's end Covid now, shall we? Step one: pound sand, Francis Collins.

The Madness Stops only when Fauci is Stopped

Originally published on December 3, 2021
in *American Thinker*.

On December 1, 2021, at a White House press conference, Tony Fauci was asked to comment about the "endgame" of Covid-19. After blabbering nonsensically and unscientifically about the pre-eminent role of his beloved — and pathetically leaky — vaccines, Fauci finally proclaimed, "I promise you that this will end."

Needless to say, St. Anthony of Vaccinia didn't deign to tell us *when* it would end, and he scurried away from the lectern even before the barrage of follow-up questions could be fully hollered.

Talk about a cliffhanger.

Anthony Fauci, the diminutive egomaniac who declared, "I represent science," the *éminence grise* whom the hapless, befuddled Joe Biden only half-jokingly referred to as our real president, will not tell us when the Covidiocy will end, because he knows the real answer: the madness stops only when he — Anthony Fauci — is stopped.

As long as Fauci has the ear of our benighted, demented president; as long as Fauci drives the multibillion-dollar endless-vaccine-by-subscription juggernaut; and as long as the amoral media and the Democrats cover for him, allowing them to punish the citizenry into total submission, the vicious lunacy that is our nation's existence under Covid will continue.

Fauci doesn't represent science any more than Bunsen Honeydew does. However, Fauci *is* the linchpin to the whole "Covid forever" enterprise. He is the cornerstone of that whole diseased, corrupt temple.

There are signs the tide is turning. BeagleGate clearly shook Fauci — as it did his protector, the biblically hypocritical NIH chief Francis Collins. However, it did so not because either one cares that their sadism was exposed, but because their phones were ringing off the hook with endless calls from angry citizens. They know we are onto them.

Rand Paul and RFK, Jr. cannot do it by themselves. Every American who wants his country back, his civil liberties back, his *life* back needs to realize that Fauci must be removed from his position atop NIAID — ideally, in an orange jumpsuit — for any return to normal to occur. People must scream this to their congressmen, their senators, to the press, to everyone who will listen, and even to those who will not.

You heard it here: in five or ten years, Fauci will be universally acknowledged as one of the most corrupt, despicable villains both in American political history and in the history of medicine. He may prove to be the most prolific purveyor of human death since Pol Pot. Read *The Real Anthony Fauci* if you think I exaggerate.

The problem is, our country cannot wait five or ten years.

Hey, Francis Collins: If you've been at the NIH too long, then what about Anthony Fauci?

Originally published on November 26, 2021, in American Thinker.

The NIH's Chief Pharisee, the monumentally hypocritical Francis Collins, recently announced his resignation from the agency, less than a month after he was called out as a liar by truth-tellers, like Professor Richard Ebright of Rutgers University, concerning the NIH's potential role in the origins of Covid-19.

On Sept. 6, 2021, Ebright, summarizing a series of evidence-filled tweets, asserted:

> [D]ocuments make it clear that assertions by the NIH Director, Francis Collins, and the NIAID Director, Anthony Fauci, that the NIH did not support gain-of-function research or potential pandemic pathogen enhancement at WIV [Wuhan Institute of Virology] are untruthful.

On Oct. 5, Collins announced his retirement as head of the NIH, effective at the end of 2021.

The proximity of Collins's resignation to the revelations of Ebright and others is obvious.

Nevertheless, Collins gave the following explanation for his retirement:

> *I fundamentally believe, however, that no single person should serve in the position too long, and that it's time to bring in a new scientist to lead the NIH into the future.*

So glad you finally saw the light, Francis. But if the twelve years you served is long enough, what about your buddy and key underling, Tony Fauci?

Fauci has been the almost unfathomably powerful head of the key NIH sub-department NIAID since 1984. *Nineteen eighty-four!*

Setting aside the Orwellian undertones that particular year evokes, note that 1984 was *37 years ago*. In 1984, 2 million Commodore 64 computers were sold, the Soviet Union and the whole Eastern Bloc were intact, and HIV had just been identified.

Hey, Francis, if more than twelve years would be too long for you, then why will Tony still be running NIAID for 37 years and counting — even after you leave?

It gets worse. Fauci has spent almost his entire adult life at NIAID. He started working there in 1968. *Nineteen sixty-eight!*

Nineteen sixty-eight, for those keeping count, was *52 years ago*. Nineteen sixty-eight was the year before Woodstock, the year before the moon landing, and the year of the Tet Offensive.

Over that half-century-plus, Fauci transformed himself from a researcher into a grant-funding potentate. According to Robert F. Kennedy, Jr.'s new book, *The Real Anthony Fauci*, Fauci exerts direct or indirect control over more than half of all global biomedical research funding. RFK Jr. calls Fauci the J. Edgar Hoover of public health.

On Oct. 20, 2021, Professor Ebright added, in what will someday be recognized as a tweet for the ages:

The NIH — specifically, Collins, Fauci, and [Lawrence A.] Tabak — lied to Congress, lied to the press, and lied to the public. Knowingly. Willfully. Brazenly.

As the pressure has mounted on Fauci to be held accountable, multiple members of the medical-industrial complex, including Collins himself, have called for critics of Fauci to be "brought to justice."

Francis Collins: If anyone on earth has been on the job too long, that person is Anthony Fauci.

And Francis: If you are anything other than the inveterate liar that Professor Ebright and others have so convincingly argued that you are, then you must do more than get out. You must take Tony Fauci with you when you go.

Fauci should 'get' while the gettin' is good

Originally published on November 9, 2021 in American Thinker.

Does Anthony Fauci see the handwriting on the wall? His boss and enabler, the immensely hypocritical Francis Collins, apparently can take a hint. He's getting out while he can. He'll ride off into the sunset with a hefty government pension and the adoration of his sycophants.

But Fauci?

He must see the storm clouds gathering. No one, not even a true dictator, can stand in front of the cameras as many times as Fauci has, can pontificate to and manipulate a nation the way Fauci has, and get away with it indefinitely. Remember the Ceacescus, Tony? Remember Colonel Gaddafi? Sure, they held on a lot longer than two years, but then, they had secret police.

Fauci has been caught repeatedly lying about his role in the coronavirus research done at the Wuhan Institute of Virology. Despite

Fauci's misdirection and obfuscation from the very start, by now everyone with a lick of sense, from Rand Paul to Jon Stewart, knows that the SARS-CoV-2 virus originated in that Fauci-funded communist Chinese lab, supported with our tax dollars. Only the willfully ignorant and terminally demented (Let's go, Brandon!) still believe otherwise.

The evidence is overwhelming that Fauci, the NIAID (his agency), and his cronies at the EcoHealth Alliance funded gain-of-function research at Wuhan, failed to follow protocols, and have lied about doing so. Fauci still denies, denies, denies, but his last remaining excuse — essentially, that the experiments he funded, *strictly speaking*, weren't *exactly*, *not quite* gain-of-function — grows more tiresome and implausible every day.

Add to this the multiple acts of unimaginable animal cruelty repeatedly perpetrated by Fauci's NIAID. The "experiments" he ran in the Tunisian desert weren't science; they were sadism. Starving sandflies to eat the faces off of beagles? Really, Tony? That's your idea of *science*? In his treatment of man's best friend, Fauci makes Michael Vick look like Doctor Doolittle. And remember: Vick did hard time.

Pro-tip, Tony: If you value your public image, don't torture Snoopy.

A vital and often ignored point about BeagleGate is that, when viewed alongside Fauci's Wuhan shenanigans, it reveals his *modus operandi*. Want to perform some nefarious experiment that would never pass institutional review in the United States? Easy! Farm it out to foreign countries where research ethics are lax or nonexistent, and pay for it with U.S. tax dollars. It's the evil scientist's equivalent of the old hobo's meme — this scientific drifter moves wherever the ethical climate suits his clothes.

Worse than BeagleGate, though less well known to the public — so far — is Fauci's role in the AIDS drug trials at the Incarnation

Children's Center in New York City in the early 1990s. The U.S. government later admitted that those trials, conducted under Fauci, done on so-called "AIDS orphans," violated federal rules to protect vulnerable research subjects. They also resulted in the deaths of dozens, and perhaps hundreds, of impoverished, orphaned minority children.

How did Fauci, America's hero two years ago, get to this point? Let's review.

First, his constant camera-seeking presence; his imperious, overbearing personality; his idiotic flip-flopping; and his never-ending falsehoods eventually wore thin. When you're a complete phony, overexposure is the kiss of death.

Second, increased scrutiny into his actions and statements surrounding Covid-19 has unearthed disturbing facts that involve him with the origins of the virus.

Third, his duplicity surrounding Covid-19, once unearthed, prompted further investigation into his past, and now the skeletons — and Anthony Fauci has a lot of skeletons — are tumbling out of his closet.

This type of narrative never ends well. Witness Andrew Cuomo.

Fauci isn't a young guy. He'll be 81 by Christmas. He's famously the highest-paid federal employee there is. Just imagine his pension. Combine that with everything he's managed to extract from the Medical-Industrial Complex all these years, and surely the guy is financially secure. He's received more awards than anyone could actually deserve, from the Presidential Medal of Freedom to the *Ordine al merito della Repubblica Italiana*. Hell, in 2020, he was named the U.S. government's Federal Employee of the Year (which, quite frankly, he ought to be, given his salary). Why should he stick around?

He can walk away now, and not face a Cuomo-esque

ending. Why doesn't he?

Does Fauci think he's riding a tiger, and he's afraid to get off? Well, he isn't, not quite yet. Even Cuomo — who, let's remember, did resign voluntarily, no matter how much pressure he was under — will ultimately skate. He'll cop a plea on the sexual misconduct charges, and off he'll go with a misdemeanor at worst.

If Fauci leaves now, he's probably in the clear. But the longer he stays on, each additional time he lies to Congress, each time he promotes yet another abusive policy, the more the hot coals accumulate over his head, and the more he pushes his luck. Someday soon, the charges are coming.

Regardless of the tradition, human wisdom proclaims the same warning to men like Fauci: hubris begets nemesis, karma's a you-know-what, and pride goeth before destruction. Always has been that way, always will. And the end result — be it Julius Caesar or Richard Nixon — is not pretty. Not for nothing, it's called the bitter end.

If Anthony Fauci had any sense, he'd resign, retire, and repent, while time and fate still allow. If not, well, he has sown the wind. Let him reap the whirlwind.

The New Law

Originally published on October 21, 2021
in *American Thinker*.

Does this sound familiar?

As a child, you are taught by those in power that you must cover your face when you go out into the world. You must do so to shield others from the seed of their destruction, which dwells within you.

Your movements are restricted as well. Some places you may go, others you may not. The manner in which you may or may not congregate with others, when you may or may not do so, and whom you may or may not see are also determined by those in charge.

You're required to submit to a medical treatment. You're told it's a very minor procedure. It's hardly a procedure at all, really. And after all, it's so important that it be done to you. In fact, it's so important that if you resist, they may hold you down and force it upon you if they have to.

Besides, what choice do you really have? You see that those who refuse to submit to all these orders are bullied, mocked, and rejected. The children cannot go to school; the adults lose their jobs. Those who refuse are second-class citizens at best and outlaws — enemies of the people, even — at worst. They are rightly and righteously ostracized as the anti-socials, the apostates, the infidels that they are.

You are taught that all of these rules, strange and harsh as they may seem, are absolutely necessary. They are for your own good. They are essential to the proper function, indeed the very survival, of society. All of your leaders tell you so. These rules are morally just. They are sanctioned from above.

This is the new law. Submit to it.

Appendix A

Open Letter to CDC Advisory Committee on Immunization Practices (ACIP) Members

Submitted to ACIP Committee Members on October 18, 2022.

October 18, 2022 – OPEN LETTER TO ACIP MEMBERS

Dear CDC Advisory Committee on Immunization Practices (ACIP) Members:

I am a practicing primary care physician with over 20 years and counting of clinical experience. I regularly counsel and advise my patients on vaccines. I am a consistent proponent of vaccines in general. I am writing to urge you NOT to place the Covid-19 vaccinations onto the CDC Child and Adolescent Vaccine Schedule.

Here is why you should NOT place the Covid-19 vaccines on the vaccine schedule:

1. **The risk/benefit ratio of the Covid-19 vaccines is UNFAVORABLE for children.** As we all now know, risk of serious illness and death from Covid is a) strongly correlated with advancing age, b) declining as the virus continues to mutate, and c) is and always has been extremely low in children. Comparison to VAERS and EudraVigilance data show there is greater risk to children from Covid-19 vaccine adverse events than from infection itself.

2. **The Covid-19 vaccines do NOT stop either contraction of the disease nor transmission to others.** As such, they are worthless in efforts to create herd immunity – unlike many other vaccines that are already on the schedule.
3. **The Covid-19 vaccines have NO long-term safety data.** How could they? They have been in use less than 2 years. Children receiving these vaccines will live for up to a century having to deal with the unforeseen risks of these therapies, the magnitude of which we have ZERO knowledge at present.
4. **The Covid-19 vaccines are much less safe than vaccines already on the schedule.** Covid-19 vaccines have well-known, well-demonstrated, significant toxicities in children and young persons (e.g. myocarditis) that occur in significant excess to other vaccines currently on the Child and Adolescent Vaccine Schedule.
5. **Mandating these extremely new and minimally tested therapies on the entire population of American school children will further undermine trust in the vaccine schedule for millions of Americans for years to come.** In short, it will backfire. Population-wide confidence in the current vaccine schedules has already been seriously undermined by the Covid-19 vaccine policies that have caused countless Americans (including thousands of health care workers) to lose their jobs. In my practice, I now see multiple patients refusing all vaccines, despite my recommendations otherwise, as a result of the coercion they felt regarding Covid-19.
6. **The prime directive of medicine is** *primum non nocere* – first, do no harm. At present, the risk/benefit ratio for the Covid-19 vaccines in children is not even close. Do

not put the blood of further, avoidable adverse events and deaths on your hands. Do NOT add the Covid-19 vaccines to the Child and Adolescent Vaccine Schedule.

Thank you. Please feel free to contact me at any time if you wish to speak further.

Sincerely yours,
CLAYTON J. BAKER, M.D.

Appendix B

Open Letter from Finger Lakes Physicians on School Reopening

Released via public press release July 13, 2020

July 13, 2020

Open Letter to Finger Lakes Region Parents, School Administrators, and Policymakers:

As parents, neighbors, and practicing physicians in the Finger Lakes Region, we have witnessed the heavy toll that school closures, "distance learning," and forced social isolation have taken on children in recent months. We submit this open letter to the community to offer evidence-based recommendations regarding the full opening of public, private, and parochial schools, that prioritize the best interests of children.

There has been a lack of communication, transparency and direction at the state, county, and local levels regarding plans to reopen schools. Scientific understanding of Covid-19 has increased tremendously in just months. Knowledge, not fear, should direct policymaking for education. This letter is intended to provide local leaders with sound recommendations for school reopening by local physicians in the Finger Lakes Region.

The current data show that Covid-19 is not a significant health threat to children.

- In children, Covid-19 infection is much less common than in adults, and rarely causes serious illness.[1,2]
- Fatalities in children are extremely rare; Covid-19 deaths in ages 5-14 make up less than 1 in 8,000 of all Covid-19 deaths in the US.[3]
- The data show that, counterintuitive though it may seem, children are significantly less efficient at spreading the virus than adults.[4,5]
- Schools have been successfully and safely reopened in numerous other countries during the pandemic.[6,7]

The prospect of distance learning presents grave academic, social, and emotional harms to children.

- Distance learning restricts every aspect of learning: Only a small, privileged population of children can flourish

---

1 Coronavirus Disease 2019 in Children — United States, February 12–April 2, 2020. MMWR. April 10, 2020 Vol. 69 No. 14, 422-6.

2 Zhang L, Peres TG, Silva MVF, Camargos P. What we know so far about Coronavirus Disease 2019 in children: A meta-analysis of 551 laboratory-confirmed cases [published online ahead of print, 2020 Jun 10]. Pediatr Pulmonol. 2020;10.1002/ppul.24869. doi:10.1002/ppul.24869

3 https://data.cdc.gov/NCHS/Provisional-COVID-19-Death-Counts-by-Sex-Age-and-S/9bhg-hcku

4 Ludvigsson JF. Children are unlikely to be the main drivers of the Covid-19 pandemic – A systematic review. Acta Paediatrica. 2020;00:1-6.

5 Leclerc QJ, Fuller NM, Knight LE *et al*. What settings have been linked to SARS-CoV-2 transmission clusters? [version 2; peer review: 2 approved] Wellcome Open Research 2020, 5:83 https://doi.org/10.12688/wellcomeopenres.15889.2

6 Reopening schools in Denmark did not worsen outbreak, data shows. Reuters, World News. May 28, 2020

7 Boffey D, Willsher K. Schools reopening has not triggered rise in Covid-19 cases, EU ministers told. The Guardian. 18 May 2020.

academically through distance learning, where academic progress is possible only for highly motivated visual learners with quality computers, high-speed internet access, and available, tech-savvy parents who can support them.
- Children who rely on the hands-on assistance of their teachers fall behind.
- Children with special needs are isolated from state-mandated services, and fall behind.
- Children from disadvantaged socioeconomic groups lose access to equalizing factors, fall behind, and the achievement gap widens.
- Children absolutely need social interactions with peers and teachers to regulate their emotions and develop social skills. Increased depression and anxiety in children have been widely reported since school closures began.[8]
- Distance learning presents an insurmountable challenge to parents who work.
- Distance learning increases parental stress for all parents. Schools provide invaluable structure and support for families. Household stress and isolation are established risk factors for child abuse. Since schools closed in the US, reports to authorities of child abuse are down - as teachers are the most frequent reporters of abuse - while hospitals have noted increased severity of child abuse cases, with more frequent child deaths, as children are

---

8 Kamanetz A. With School Buildings Closed, Children's Mental Health Is Suffering. NPR. May 14, 2020.

trapped at home with abusers.[9]

We have referred to guidelines from leading medical organizations, particularly the American Academy of Pediatrics (AAP). We have reviewed guidelines released by other US state governments. We have reviewed the scientific data on Covid-19 and children in detail, including data obtained from many countries that have already successfully reopened their schools. We have spoken with many parents and children in our community. We recommend:

It should be *given* that our children return to full day, every day, in-school instruction at the start of the school year. The AAP correctly states that "all policy considerations for the coming school year should start with the goal of having students physically present in school."[10] Given what we now know – how safe school openings have proven to be, and how much school closings do indeed harm children – it is wrong to needlessly subject our children to further educational, psychological, and physical harm by denying them in-school education.

In-school elementary and secondary education should be officially categorized as essential services. In New York State, liquor stores were deemed essential from the start of the Covid-19 lockdown. It is unconscionable that in-school learning for our children is not.

---

9 Schmidt S, Hatanson H. With kids stuck at home, ER doctors see more severe cases of child abuse. Washington Post, April 30, 2020.

10 https://services.aap.org/en/pages/2019-novel-coronavirus-covid-19-infections/clinical-guidance/covid-19-planning-considerations-return-to-in-person-education-in-schools/

Decisions regarding the details of school reopenings in Upstate New York should be made locally. The severity of the pandemic varies dramatically across the State. Logistics also vary widely, sometimes even between neighboring school districts. It is nonsensical that blanket decisions made in Albany are applied simultaneously to Schuyler County (15 total cases, 0 deaths), Queens County (65,568 total cases, 5,048 deaths),[11] and all counties in between.

The AAP guidelines should be a primary source of guidance for reopening. These are reasonable, evidence-based guidelines that prioritize the needs and best interests of our children. The data clearly show that children are at low risk due to Covid-19, and do not spread it efficiently to others. Safety measures should focus on protecting vulnerable adults.

1 meter social distancing between students is a reasonable standard. The World Health Organization, AAP, other US states, and many countries that have reopened schools accept this standard.[12,13] Considerable evidence shows that 1 meter provides the benefits of

---

11 COVID NY Deaths by County. NYS Department of Health Website, accessed 7-2-20.

12 https://www.npr.org/2020/07/02/886845449/massachusetts-education-commissioner-on-states-plan-to-reopen-schools-in-the-fall

13 Kelland K. One meter or two? How social distancing affects COVID-19 risk. Reuters, Health News,
June 23, 2020

distancing.[14,15] It lessens logistical barriers to full school attendance created by longer distancing requirements. Additional measures such as masking can be added to further reduce risk.[16]

The rational approach is to protect and accommodate vulnerable persons while keeping schools open, not to harm the entire population of children by closing schools. Practical, evidence-based accommodations for vulnerable individuals can and should be made, as well as for children whose parents are not yet comfortable with a return to school.

Athletics, arts, and extracurricular activities should be resumed to the fullest extent possible. These activities provide children with vital social, psychological, and physical benefits. The National Federation of State High School Associations (NFHS) Sports Medicine Advisory Committee (SMAC) has released guidelines that can serve as a reasonable starting point for athletics. Private and club sports have already reopened in New York. It is nonsensical that equivalent school-based athletics and activities should remain barred.

---

14 Chu DK, Akl EA, Duda S, et al. Physical distancing, face masks, and eye protection to prevent person-to-person transmission of SARS-CoV-2 and COVID-19: a systematic review and meta-analysis. The Lancet 2020 doi: 10.1016/S0140-6736(20)31142-9; 0810.1016/S0140-6736(20)31142-9.

15 Qureshi Z, Jones N, Temple R, Larwood J, Greenhalgh T, Bourouiba L. What is the evidence to support the 2-metre social distancing rule to reduce COVID-19 transmission? CEBM, Oxford, June 22, 2020.

16 Chu DK, see above.

The successful reopening of schools requires a mature and courageous analysis of risks and benefits, based on our growing knowledge of Covid-19. Parents should understand the very small risk Covid-19 poses to their children, yet also recognize the concerns of older adults. Teachers must understand the finite risk to themselves and take appropriate measures, while realizing the risk they face is less than that of health care workers, and less than workers in a number of other industries. Safety measures already successfully used in other countries can be employed to protect teachers. Administrators and politicians must realize that an excessively risk-averse strategy will do great harm to students, and might ultimately result in greater legal problems than would a sensible, evidence-based reopening.

Our children deserve better than they are getting from us now. We now know enough about Covid-19 to safely reopen schools and help return our children's lives and education to normal. This has already been done in dozens of other countries. We must develop the necessary courage and determination to do so here.

Respectfully submitted,
Clayton J. Baker, M.D. (Author)

Co-signed by:
Lynette Froula, MD
Todd Gerwig, DO
Elizabeth Feltner, MD
Catherine C. Tan, MD
Nathan Ritter, MD
Gregory Finkbeiner, MD
Darren Tabechian, MD
Jill Reidy, MD

Catherine Przystal, MD
Sasha Nelson, MR
Raymond Tan, MD
Beth A. Ritter, MD
Heather Sobel, MD
Ann Olzinski-Kunze, MD
Stephen Mawn, MD, JD, MPH
Sara Connelly, MD
Robert C. Block, MD, MPH
Gregory Kunze, MD
Karen C. Mawn, MD
Megan J. Rashid, MD
Hani H. Rashid, MD
Sareena Fazili, MD
Jamil Mroueh, MD

Acknowledgements

George Orwell wrote that "writing a book is a horrible, exhausting struggle, like a long bout with some painful illness." Considering this book largely concerns Covid, that quote takes on extra meaning for me. Suffice it to say that this volume would not exist if I had to do it all by myself. Many thanks are in order.

First, I thank my family. To my mother and father I owe everything. I deeply thank my wife Jill, whose love, constancy, and loyalty over the last thirty years have been the foundation upon which our life, our careers, and our family have been built. Her patience and forbearance throughout the last five years held all those things together.

I thank my children, who have been courageous and humane beyond any reasonable expectation during an historically difficult time, and who have never failed to encourage me. Above all else, they and their generation are the reason I write.

I thank all the people at Brownstone Institute for their continuing support and interest in my work, and for their help in producing this volume. Jeffrey Tucker, Lou Eastman, and their team created a unique and powerful place for independent writers and thinkers to survive at a perilous time in history for free speech and honest inquiry. I am honored to be part of it.

I am grateful to the editors at other media outlets that have published my work, including *American Thinker*, Children's Health Defense, *Epoch Times*, *RealClear Health*, *Daily Clout*, Minding the Campus, Eagle Forum, and others.

A thank you is due to my co-authors for several of the essays here: Brian Hooker, Heather Ray, and Joshua Stylman.

Thanks to all of the extraordinary dissidents, refuseniks, and contrarians that I have had the great honor to meet and struggle alongside over the past five years. There are too many of you to name; hopefully there are enough of you to prevail. I hope and pray my writing adds a few drops to the oceans of work you continue to do to preserve human health and liberty.

A special thank you goes to my readers. There is nothing more gratifying to a writer than to know that others read what you write.

Above all, I thank God for inviting me to the banquet.

About Brownstone Institute

Brownstone Institute, established May 2021, is a publisher and research institute that places the highest value on the voluntary interaction of individuals and groups while minimizing the use of violence and force, including that which is exercised by public authority.

Published by Brownstone Institute
Austin, Texas

Index

A

Abe, Shinzo, 128
Abusive relationship, WHO, 115–120
Adams, Sam, 34
Adverse reactions, mRNA VINOs, 162
Advisory Committee on Immunization Practices (ACIP), 251–253
American Academy of Pediatrics (AAP) guidelines, 258, 259
American Animal Hospital Association Guidelines, 51
Annenberg Foundation, 130
Arlen, Harold, 236
Arnaud, Annie, 100
Assange, Julian, 3
AstraZeneca, 126, 185
Auci, Fantoni, 57
Audenz, 196
Autoimmune diseases, 135
Autonomy, medical ethics, 12–13
　confidentiality, 15
　defined, 13
　informed consent, 13–15
　protection against coercion, 17–18
　truth-telling, 15–17
Avian influenza. *See* Bird flu (H5N1)

B

Bailey, kennel cough, 49–53
Baldwin, Tammy, 155
Bancel, Stephane, 95
Bates, Gil, 55–56
Bayh-Dole Act of 1980, 182–184
Bell, David, 81–82, 117, 119
Beneficence, medical ethics, 18–20, 73
Bhattacharya, Jay, 111–112
Biden, Joe, 78, 119, 202, 233, 239
　dementia, 203–207

Big Medicine, 28, 140, 227
Big Pharma, 137, 140, 181–182, 229
　Bayh-Dole Act of 1980, 182–184
　direct-to-consumer advertising, 86–87, 188–189
　encode medical freedom, American law, 189–191
　fact-checking, 128
　NCVIA of 1986, 184–186
　PREP Act of 2005, 187–188
　Project Bioshield Act of 2004, 186–187
Bill of Rights, 70, 71, 75, 189–190, 234
Biological Weapons Convention of 1975, 144, 171
Bioweapons, 158
Bird flu (H5N1), 148, 167–168
　Fouchier, R.A.M. (Ron), 149–150
　Kawaoka, Yoshihiro, 148–149
　Marrazzo and Ison, 193–197
　reasons, 168–169
　in United States, 150–152
Birx, Deborah ("Scarf Lady"), 137, 152, 154, 167, 195, 196
Black Death, 58
Blatant coercion, 14
Bloomberg News, 122, 123
Body piercings, 135
Bomb, 58
Bourla, Albert, 7, 19, 49, 52, 228–229
Bowden, Mary Talley, 80, 94
Brayne, Carol, 205
Bulger, Whitey, 110
Bush, George W., 186, 187

C

Caligula, 199–202
Cardozo, Benjamin, 13

Carter, Jimmy, 182
Cartland, David, 111
Caveat emptor, 28
Chen, Clifford, 200
Child and Adolescent Immunization Schedule, 209
   American parents, 212
   herd immunity, 210
   natural immunity, 210
   pediatric Covid-19 vaccine programs, 211
   prime directive of medicine, 213
   risk/benefit ratio, 210
   safety data, 210–211
   schools, CDC schedule, 211–212
   trusts, immunization schedule, 212–213
Chilton, Alex, 7
Chronic atrial fibrillation, 125
Cochrane, 42, 46
Coercion, 14, 73
Cohen, Mandy, 75
Collins, Francis, 16, 234, 240
   Bayh-Dole Act of 1980, 184
   and Ebright, Richard, 236, 241–243
   farewell address, 235–237
   and Fauci, Anthony, 112, 242–243
   Great Barrington Declaration, 217, 228
   retirement, explanation, 242
Communist Chinese Party, 232
Confidentiality, 15, 72
Covid-19 boosters, health care students
   clinical sites, 105
   colleges and universities, 103–104
   dangers of, 104
   medical care, 106–107
   No College Mandates, 105–106
   unnecessary risks, 106
Cox, Bobbie Anne, 87–88
Coyle, John, 105
Cummins, Ivor, 118
Cuomo, Andrew, 25, 226, 232, 247, 248
Cuomo, Fredo, 226
Cuomo, Sonny, 226

D

De Blasio, Bill, 233
Declaration of Independence, 70, 71
Dementia, Biden's, 203–207
Democratic Republic of the Congo (DRC), 175, 176
Derelicht, Saul, 56
DeSantis, Ron, 68, 78, 81
Diabetes, 135–136, 138
Digital Services Act, 122
Dinh, Thi Thuy Van, 81–82, 117, 119
Direct-to-consumer advertising, 86–87, 188–189
Disease X, 172
Dissidents, Covid, 3–4
Distance learning, 256–258
Djokovic, Novak, 18
Dostoevsky, Fyodor, 6, 9
Dump, T. Ronald, 59–60
Durbin, Dick, 184

E

Ebright, Richard, 236, 241–243
Eisenhower, Dwight David, 92–93, 96
Emergency Use Authorizations (EUAs), 163, 186, 187, 233
Evers, Tony, 155

F

Fact-checking, 128–131

Farmer's protests, 97–101
Fatal defiance, 131–132
Fatal toxicities, 60
Fauci, Anthony, 1, 16, 19, 24, 94, 228, 230, 233, 239–240, 248
   AIDS drug trials, 246–247
   awards, 247
   Bancel, Stephane, 139
   empathy, 112
   experiments, Tunisian desert, 246
   gain-of-function research, 81
   and Leonhardt, David, 223, 224
   misdirection and obfuscation, 246
   NIAID, 147, 176–178, 193, 232, 235, 240, 242–243, 246
   official lies, 15
   Omicron variant, 234
   role in coronavirus research, 245–246
   US tax dollars, 95
   West Nile virus, 167–168
"Fear porn", 152–154, 168
Fetterman, John, 199–202
Fico, Robert, 121–123
Fire Department, 141–143
See also Pandemic preparedness
Fisman, David, 113
Fouchier, R.A.M. (Ron), 149–150, 196
The Foundation for Vaccine Research, 149
Four pillars of medical ethics
   autonomy, 12–18
   beneficence, 18–20
   justice, 25–26
   non-maleficence, 20–24
   overview, 11–12
   solutions and reform, 27–28

G
Gain-of-function research, 81, 144, 151, 193, 196, 246
   Fouchier, R.A.M. (Ron), 149–150
   H5N1 Bird flu, 151, 154, 197, 198
   International Bioweapons Convention (1975), 155
   monkeypox virus, 171–179
Garland, Judy, 236
Gates, Bill, 1, 8, 83, 117, 124, 130, 229, 233, 234
Ghebreyesus, Tedros Adhanom, 1, 82, 124
Ghoula, Adalbert, 57, 59
Ginsberg, Allen, 109–110
Global Deterioration Scale (GDS), 205–206
Gold, Simone, 94
Grady, Christine, 22
Great Barrington Declaration, 111, 112, 217, 228
The Guardian (newspaper), 122

H
H5N1 influenza virus. *See* Bird flu (H5N1)
Hamilton, Margaret, 227
Harris, Kamala, 203–204
Hippocrates' *Epidemics*, 20–21
Hochul, Kathy, 87–88, 212, 213, 217, 226–227, 233
Hoff, Jimmy, 226
Hoover, J. Edgar, 243
House Committee Report, 177, 178
Howl (Ginsberg), 109–110
Human Papilloma Virus (HPV) vaccination, 161
Humor, 6
Hunter, Robert, 110

Hutson, Christina, 173
Huxley, Aldous, 31

I
IHR. *See* International Health Regulations (IHR)
Immune system, 139
Incompetent patients, 14
Informed consent, 13–15, 72
Injection hormonal medications, 135
Insulin, 135, 138
International Bioweapons Convention (1975), 155
International Bird Flu Summit, 169
International Health Regulations (IHR), 81–83, 121–122, 132
International Monetary Fund (IMF), 127
Ison, Michael G., 193–197

J
Jefferson, Thomas, 190
Jensen, Scott, 94
Johnson, Boris, 26
Johnson, Ron, 155
Johnson & Johnson, 130, 147, 185
Jones, Brea, 129–130
Jordan, Michael, 148
Justice, medical ethics, 25–26, 73
Juvenal, 31

K
Kaczynski, Ted, 110
Kaur, Kulvinder, 111–113
Kawaoka, Yoshihiro, 148–149, 168, 196
Kekatos, Mary, 153
Kelly, Megyn, 78
Kennedy, Robert F., 78, 81, 217, 230, 240, 242–243

Kennel cough, 49–53
Kesey, Ken, 110
Klain, Ron, 201
Klar, John, 161–162
Kogon, Robert, 98–99
Kulldorff, Martin, 111

L
Leonhardt, David, 221–224
Levine, Rachel, 25–26
Lindley, Kat, 100, 119
Lockdowns, 8, 22, 87
 Abe, Shinzo, 128
 Bill of Rights, 70
 burdens of, 26
 Communist Chinese Party, 232
 Kaur, Kulvinder, 111–112
 liquor stores, 31, 258
 Nkurunziza, Pierre, 126
 public health emergency, 216
 Trump presidency, 59–60
Lowenthal, Andrew, 80
Lyme disease, 134
Lysenkoism, 218

M
Macron, Emmanuel, 98
Magufuli, John, 124–125, 129
Malaria, 134, 176
Malik, Ahmad, 111
Malone, Robert, 229
Manson, Charles, 110
Marrazzo, Jeanne, 193–197
Martial law, 94, 216
Masks, 45
 controlling people, 47–4
 controlling virus, 46–47
Massie, Thomas, 162, 185

Mature minor doctrine, 161
McCartney, Paul, 66
McCullough, Peter, 83, 94, 111, 216, 217
McVeighs, Timothy, 70
Mechanisms of harm, 8
Medical freedom, 65
  beneficence, 73
  censorship, silencing, intimidation and punishment, 74
  characterization of, 67
  coercion, 73
  confidentiality, 72
  definition of, 66, 70–72
  informed consent, 72
  justice, 73
  Medical Freedom Party, 69
  the Nation (magazine), 67–68
  new year's resolutions, 77–89
  non-maleficence, 73
  open and honest debate, 73–74
  outside influences, 74
  patient autonomy, 72
  patient-physician partnership, 74
  patient redress, 74
  protocols, 74
  public health directives, 73
  public health officials, 75
  refusal of treatment, 73
  Senate Bill 252, 68
  truth-telling, 73
  the Washington Post (newspaper), 66–67
The Medical Freedom Hustle, 67
The Medical Freedom State, 68
Medical humanities and bioethics, 5
Military medicine
  bioweapon, 95
  civil liberties, 94
  corporations, 91–92
  Eisenhower's farewell speech, 92–93, 96
  ground-level personnel, 92
  life-or-death issues, 92
  medical-industrial complex, 93, 95
  SARS CoV-2 virus, 95
  top-down diktats, 94
Minor consent, 161
Moderna, 84, 126, 137, 162, 163, 185
Modus operandi, 3, 246
Moïse, Jovenel, 125–126
Monkeypox (Mpox) virus, 167, 172
  case-fatality rates, 174–175
  deaths, 175
  Democratic Republic of the Congo (DRC), 175, 176
  Fauci's NIAID, 176–178
  SARS CoV-2 virus, 174
  size of, 174
  in US, 172–173
Moss, Bernard, 177
mRNA vaccines, 18–20, 42, 79, 84, 185, 228
  adverse events and deaths, 160
  Big Pharma and Moderna, 137
  development and administration, 39–40
  moratorium, 84–85
  removing from market, 83–84
  RSV injection, 138
  VINOs, 162–164

N

Nader, Ralph, 187
Nakhlawi, Razzan, 67
Nass, Meryl, 3, 111, 119
The Nation (magazine), 67–68

National Childhood Vaccine Injury Act (NCVIA) of 1986, 38, 41, 85, 136, 184–186
National Domestic Violence Hotline, 116
National Institute of Allergy and Infectious Disease (NIAID)
   Fauci, Anthony, 147, 176–178, 193, 232, 235, 240, 242–243, 246
   Marrazzo and Ison, 193–197
Ndayishimiye, Evariste, 127
Needle penetration, 140
Neutron bomb, 58
New law, 249–250
Newsom, Gavin, 26, 212
New year's resolutions, medical freedom, 77–78
   Covid-19 mRNA vaccines, 83–84
   direct-to-consumer advertising, pharmaceuticals, 86–87
   gain-of-function research, 81
   moratorium, mRNA platform, 84–85
   NCVIA of 1986, 85
   play offense, 87–89
   politicians, 80
   truthful narratives, 79–80
   vaccine mandates, 85–86
   WHO, 81–83
The New York Times (newspaper), 122, 204, 221
NIAID. *See* National Institute of Allergy and Infectious Disease (NIAID)
Nkurunziza, Pierre, 126–127, 129
Non-maleficence, medical ethics, 20–24, 73

O
Obama, Barack, 196
Offit, Paul, 150
Omicron variant, 233–234
Open letter
   to ACIP members, 251–253
   Finger Lakes physicians on school reopening, 255–261
   to students and parents, 157–166
Operation Warp Speed, 163
Orient, Jane, 154, 155
Oropouche virus, 168
Orwell, George, 31
Osorio, Jorge, 173
The Overpopulation Bomb (Derelicht), 56
Oz, Mehmet, 200

P
Pan-Africa Epidemic and Pandemic Working Group, 131–132
Pandemic preparedness, 141–142
   Bird flu in United States, 150–152
   Covid injections, 147
   "fear porn", 152–154
   Fouchier, R.A.M. (Ron), 149–150
   gain-of-function research, 144
   Kawaoka, Yoshihiro, 148–149
   monkeypox (Mpox) virus, 171–179
   multi-step process, 143–144
   pathogens, interventions, 145
   recommendations, 154–156
   risk factors, 142
   SARS-CoV-2, 147–148
   taxpayer funds, 142
   VAERS, 146–147
Patent and Trademark Law Amendments Act (Public Law 96-517). *See* Bayh-Dole Act of 1980
Patient autonomy, 17, 72
Paul, Rand, 240, 246

Paul, Ron, 66
Paxton, Ken, 88
Pelosi, Nancy, 26, 118
Pfizer, 83, 88, 93, 162, 163, 181, 185
Plague, 58
Plotkin, Stanley, 150
Politella v. Windham, 161–162
PolitiFact, 130
Prime directive of medicine, 213, 252–253
Project Bioshield Act of 2004, 186–187
Project MK-Ultra, 110
Public Health™, 45, 47–48
Public Readiness and Emergency Preparedness Act (PREP Act) of 2005, 187–188

R
Ramaswamy, Vivek, 78
Reagan, Ronald, 136, 185
The Real Anthony Fauci (Kennedy), 217, 230, 240, 242
Redfield, Robert, 153–154, 167
Re-sorting process, 4
Respiratory Syncytial Virus (RSV) injection, 138
Respiratory virus, 59, 60, 63
Reverse Transcriptase-PCR testing, 150
Rice, Susan, 201
Risch, Harvey, 46
Risk/benefit ratio, Covid-19 vaccines, 210, 251
Robert Wood Johnson Foundation, 130
Rogers, Toby, 183
Rothman, Jay, 149, 156
Rush, Benjamin, 190

S
Sadistic streak, 23

Safety signals, 60
SARS-CoV-2 virus, 79, 95, 154, 158, 174, 224, 246
  creation of, 147–148
  mask mandates, 46
  mild illness, children, 19
Schlafly, Phyllis, 187
Schlob, Kraut, 56
Schwab, Klaus, 1
Semaglutide, weight loss treatment, 136
Senate Bill 252, 68
Seven health policy habits, insanely progressive people, 215–216
  government control, medicine, 218
  government-pharmaceutical-industrial complex, 217
  Lysenkoism, 218
  mandatory jab policies, 216
  martial law, 216
  physicians, scientists, and health care workers, 216–217
  predatory medical and social policies, 217
Sloth virus, 168
Smith, James, 129
Snowden, Edward, 3
Social distancing, 126, 127, 259–260
Social media platforms, 234
Soft-core totalitarianism
  civil rights, 31–32
  defined, 29–30
  dystopian narrative, 32–34
  remedies, 34
  semi-anesthetized/semi-tolerable dystopia, 30–31
  soft-core pornography, 30
"Spoke" protein, 61
Sterilization, 58–59

Stevens-Johnson Syndrome, 140
Stewart, Jon, 246
Students and parents, open letter to, 157–158
  college and health care studies, 159–160
  elementary, middle, and high school student, 161–162
  learn to say "no", 165
  medical treatment, 165
  mRNA and vaccine pipelines, 166
  problems and reasons, 159
  purposes, 164
  safety concerns, mRNA VINOs, 162–164
  self-protection, 166
Subtler forms, coercion, 14
Sudden death, 61, 127, 131
Suetonius, 199
Suluhu Hassan, Samia, 125

T
Tattoos, 135
Thimerosal, 137
True heroes, 3
Trump, Donald, 78, 128
Truth-telling, 15–17, 73
Tucker, Jeffrey, 6
Turbo cancers, 62, 63
Type-1 diabetes, 138
Type-2 diabetes, 135

U
USDA Southeast Poultry Research Laboratory (SEPRL), 151, 152, 195–196

V
Vaccine Adverse Event Reporting System (VAERS), 40, 42, 146–147, 160, 185, 233
Vaccines, 35
  anti-vaxxers, 35, 36
  CDC Child and Adolescent Immunization Schedule, 39
  definition of, 37–38, 40
  Federal law, 41–42
  medical treatments, 40–41
  NCVIA of 1986, 38, 41
  pathophysiological mechanisms, 38
  pharmaceutical industry, 39
  public investigation, 42
  reviews, 42
  role of, 43
  safe and effective, 36
  safety signals, 60
  -targeted diseases, 21
  toxicity of, 38
  trouble with, 41–43
  See also mRNA vaccines
Vaccines-In-Name-Only (VINOs), 162–164, 185
VAERS. *See* Vaccine Adverse Event Reporting System (VAERS)
Vaxteen, 161
Vick, Michael, 246
VINOs. *See* Vaccines-In-Name-Only (VINOs)
Virchow, Rudolf, 47
Vulnerable populations, 17

W
Wakefield, Andrew, 3
Walensky, Rochelle, 16–17, 19, 210, 227
War, 57–58

Washington, George, 190
The Washington Post (newspaper), 66–67
Weingarten, Randi, 226
Weitz, Lori, 8
Welby, Marcus, 230
Wen, Leana, 195
West Nile virus, 167–168
Whitmer, Gretchen, 212, 213
Wigington, Dane, 3
World Enslavement Forum, 56–57
World Health Organization (WHO)
    Abe, Shinzo, 128
    abusive relationship, 115–120
    COVAX program, 126
    fear porn, 153
    Fico, assassination attempt on, 121–123
    IHR, 81–83, 121–122
    Magufuli, John, 124
    monkeypox outbreak, 174–175
    Nkurunziza, Pierre, 127
    pandemic proposals, 117, 118
    Pandemic Treaty, 131–132

X
Xi Jinping, 230

Z
Zuckerberg, Mark, 229

Made in the USA
Monee, IL
01 May 2025

16667932R00173